Everything I Want To Do
Is Illegal

Everything I Want To Do Is Illegal

by
Joel Salatin

Polyface Inc.
Swoope, Virginia

This publication is designed to provide accurate and authoritative information in regard to the subject matter covered. It is sold with the understanding that the publisher is not engaged in rendering legal, accounting or other professional service. If legal advice or other expert assistance is required, the services of a competent professional person should be sought. *From a declaration of principles jointly adopted by a committee of the American Bar Association and a committee of publishers.*

Library of Congress Control Number: 2007928608

ISBN: 0-9638109-5-2

New ISBN *(effective 1/1/2007)***: 978-0-9638109-5-3**

Other books

by

Joel Salatin

Pastured Poultry Profit$
Salad Bar Beef
You Can Farm
Family Friendly Farming
Holy Cows & Hog Heaven

All books available from:

Acres USA	1-800-355-5313
Amazon.com	www.amazon.com
Chelsea Green Publishing	1-800-311-2263
Stockman Grass Farmer	1-800-748-9808

or by special order from your local bookstore.

Contents

Acknowledgments...xi
Foreword...xii
Introduction...xv

The Past
Ch. 1 - The Original Essay 2
Ch. 2 - Raw Milk and Dairy...12
Ch. 3 - PL 90-492...28
Ch. 4 - Custom Beef 50
Ch. 5 - Bacon...66
Ch. 6 - Salmonella..80
Ch. 7 - 1,000,000 Mile Chicken 90

The Present
Ch. 8 - Organic Certification 106
Ch. 9 - Government Grants - Best Management Practices.......122
Ch. 10 - Government Grants - Conservation Easements 136
Ch. 11 - Restaurants 146
Ch. 12 - Predators and Endangered Species.......................158
Ch. 13 - Sawmills Are Out 172
Ch. 14 - Zoning 182
Ch. 15 - Labor 196
Ch. 16 - Housing 214
Ch. 17 - Insurance 234
Ch. 18 - Taxes 250

The Future
Ch. 19 - Avian Influenza..264
Ch. 20 - Bioterrorism...278
Ch. 21 - National Animal Identification System.................286
Ch. 22 - Mad Cow..308
Ch. 23 - Animal Welfare...320
Ch. 24 - Options..332
Index...348

Acknowledgments

Thank you, lovely bride of 26 years, Teresa, for encouraging me and sharing this farm passion with me. Thank you for your unconditional support, your brainstorming, and your righteous indignation about the things that deserved contempt.

I thank my mother Lucille, daughter Rachel, son Daniel, and daughter-in-law Sheri for helping keep me sane in an insane world. Dad, I wish you could see this book. You'd love it. Thanks for all the political discussions growing up. This apple didn't fall far from the tree.

Thank you apprentices Nathan Vergin, Jordan Winters and Matt Rales for picking up the slack during the winter of 2006-7 so I could stay inside and plug away at this project.

Thank you friend and co-conspirator Jeff Ishee for your consistent desktop publishing miracle. And for being the quintessential encourager.

Thank you Deborah Stockton and Sheri for proofing the original manuscript. I wonder how many things we all missed.

Thank you Robin Leist for taking Rachel's cover idea and really making it work. It's a beautiful job.

Thank you grandsons Travis (3) and Andrew (1) for lighting my life anew. May your world be better than mine.

Thank you Polyface patrons for sticking with us, for not caring about labels, inspections, or any of that other stuff. You make our world better. Bless you all.

FOREWORD

Historians tell us that Talleyrand once told Napoleon, "You can do anything with bayonets, Sire, except sit on them." In the language of the court this meant that obedience to the law ultimately depended on cooperation, not force.

In this marvelous little volume, farmer, journalist, philosopher Joel Salatin makes a good case for withholding that cooperation. Without once saying so, he seems to tell us, "Run those solipsistic dweebs out of budget."

Salatin's journal is truly a handbook for young people who have a passion for the farm. Like Prince Hal, Joel Salatin achieved a maturity capable of crowding his adversaries into the fatal flaw of contradiction.

He sucked up rural wisdom like a sponge, first from his progenitors, then by plugging in a rare commodity -- uncommon good sense. It told him that the greatest impediment to the resettlement of America was capitalization, a Gordian knot that not even a hostile bureaucracy has been able to unravel entirely.

Salatin's capitalization was started by granddad, continued by dad, then by Joel's own prudent upbringing. Capitalize on savings, expand on earnings, cut costs to the bone, then watch the doors fly open. Joel and his wife shared a two-apartment home with his parents for seven years to reach freedom's launching pad.

But there's much more to the story. In the real world, he found that power obsessed otherwise unemployable rent-a-cops, who treated farmers -- small-scale and large -- like fungus on the left small toe. Soon enough, these types became instant lawyers and judges, grammarians and Aristotelean logicians. Precedents meant nothing to these automatons with their every-ready Eichmann defense. The way these come-lately Javerts saw it, they were the law because they said so.

As Joel Salatin walks the reader through workaday reality, it becomes transparently obvious that the character Peter Stuyvesant in Knickerbocker Holiday was right: "Governments, one and all, partake of the nature of rackets. They become partners in crime and ultimately annihilate the civilization over which they preside."

Sell a chicken, a pound of beef, fresh milk, and chances are you're performing an illegal act. I recall a TV program of a few years back in which the fellow was raising cattle, growing feed grains, and literally starving. You say, "There's something wrong with that picture. Why didn't he slaughter a beef cow and spare his wife the humiliation of standing in a food commodity line?" Well, it's illegal. He can't slaughter a critter on his own farm.

Everything is illegal now unless you pick up a pail of permits, licenses, certificates and nods of approval from a bureaucrat who doesn't know the difference between bleeding a chicken and condemning the meat to mediocrity by killing with electric shock.

Even those compromised by the IQ inhibitors in their junk food know that real food contamination takes place in CAFOs, huge slaughter houses, massive processing plants and warehouse storage. As a consequence, the federal and state governments single out small farmers, local food production and local sales facilities for compliance with draconian measures.

For a real lesson in social science and mendacity, read these pages. I am usually slow to anger, but honest to God, I fantasize about that legendary steer:

His far hind leg in Buffalo,
His front one in Tacoma,
His left rear half in Jacksonville,
His front one in Pomona.
And after that steer et all that hay
Wouldn't there be a fuss,
If what that steer did to Washington
Was what Washington did to us?

This much said, I continue to wonder aloud what stretch of the imagination caused organic growers to ask Washington for an Organic Standards Act. It had "blunder" written all over it from the start, and Joel Salatin explains why. He explains the NAIS oppression coming up, the war on fresh milk, on sound education -- in a word, on freedom.

I have recently been asked to name 100 books essential to a good eco-farming library. I would name <u>Everything I Want to Do Is Illegal</u> as numero uno. All the knowledge about cation exchange capacity, the carbon cycle, nitrogen uptake, and the rubrics of working with nature mean nothing unless one understands how the bureau people cold-deck him or her in the card game of life.

Cold-decked or not, there is power in numbers. Gandhi told us this. "How can a few thousand Brits control millions unless we comply? We now propose to withhold compliance!" The words are approximate; the thought is not.

Salatin suggests more than a wave of discontent. As consumers come to the aid of their suppliers of bio-correct food, numbers can discourage the bureaucrats and defuse their Eichmann excuses.

This book is only partially an angry testament. It is full of Mark Twain humor, and it surely contains a bit of Joseph Heller's Catch 22, and Milo Minderbinder to boot -- Milo, the crafty scofflaw who, like many bio-correct producers, elude their own Catch 22s.

-- Charles Walters

INTRODUCTION

"But is it legal?" This is by far and away the most common question I am asked after doing a workshop on local food systems and profitable farming principles. My blood boils every time that happens. Not at the fearful farmer, but at the system that thinks we're a successful culture because we have more prisoners in America than farmers. To applaud ourselves for such a statistic is despicable.

Would-be local food farmers literally spend their days looking over their shoulders wondering what bureaucrat will assault them next. And yet, what could be more noble, more right, more good than neighbor-to-neighbor food sales?

If a little girl wants to make cornbread muffins and sell them to families in her church, why should the first question be "but is it legal?" As a culture, we should praise such self-motivated entrepreneurism. We should be presenting her with awards and writing stories about her creativity.

Our farm, Polyface (the Farm of Many Faces) has been featured in countless publications and media. Most recently, we starred in the New York Times runaway bestseller <u>Omnivore's Dilemma</u> by author extraordinaire Michael Pollan. All this notoriety has vaulted our family farm into the spotlight, the darling of local food advocates around the world, poster children of artisanal foods. Indeed, Pollan would never have written about us had we shipped him a grass-finished steak. That is how serious we are about local and bioregional food systems.

What many people do not understand, however, is that at every step on this journey toward success, government officials have unceasingly tried to criminalize us, demonize us, dismiss us, and laugh at us. We have fought, clawed, cried, prayed, argued and threatened. The point is that if it had been up to public servants, Polyface would not exist. And the struggle is not over. Some battles, as you will see, we did not win. Some we refuse to fight. The war goes on.

My heart breaks for others who did not start as early as we did (1961) on this local food journey, or who are not as legal savvy, who get routinely pummeled by these government officials. Many give up. Survivors emerge battered and frustrated; often angry. And justifiably so.

Supporters of local, heritage, artisanal, organic, ecological, sustainable, humane, biodynamic food need to know that every day, their food farmer friends receive visits, phone calls, threats, summonses, confiscation, and criminal charges. The harassment from government officials would make your hair stand on end. This book is about one such farmer's lifetime of dealing with these issues. Real stories. Real thoughts.

I am not an attorney. Do not expect this book to offer legal advice. In fact, I'm sure some aspects may be technically incorrect. Or my perception may be incorrect. But my perception is my reality. The fact is that if I believe it's illegal, it affects my decisions. Most farmers won't spend $500 to get an attorney's counsel for these things. Besides, most attorneys have no clue because they don't know about these issues.

What I have tried to do is lay out, as accurately as possible, my side of these stories. I have purposely stayed away from legal minutiae in order to make it more enjoyable to read. If I have overstated something or missed a point, it is an oversight and not maliciously or consciously intended. I have also purposely stayed away from similar incidents involving friends and acquaintances. Some are currently being litigated or otherwise negotiated. The average person would not believe the things going on out here in the countryside. They are horrendous. But dumb small farmers don't make the news much.

I know that many small businesses deal with similar issues, but my background is farming and that is the context of my stories. And it's a good context, because what could be more basic in any culture than its food? If this is the way we treat our food producers, heaven help the rest of the small businesses.

I'm sure some people who know how upbeat and optimistic I am will think this book springs from anger and bitterness. Anger yes; bitterness no. But I do think anger aimed at evil is a good thing. I do not think it helps any culture to dismiss elements that deprive the populace of righteousness.

I hope that this book will awaken deep within your soul a righteous indignation against the entrenched political-industrial-bureaucratic food fraternity and a deep love for farmer-healers who love their land, plants, animals, and patrons. Each emotion is necessary for balance.

Another reason I wrote this book is so that my grandchildren will know their legacy. I don't know if their farming world will be easier or harder than mine. Much depends on how this slug-fest between the powerful industrial forces and the grassroots local food movement turns out. Armed with this book, I hope our side will become more passionate and articulate in this struggle. And ultimately prevail.

But I wanted to set the record down, in black and white, to preserve the stories, preserve the struggle, preserve the history. As I write this with tears running down my cheeks, thinking of those little guys growing up into a world more centralized, more globally-oriented, more Wal-Martized, I want them to know what was and what could be. I want them to catch a vision of a righteous food system, a healing agrarianism, a local farm food ministry. May it never vanish from the earth.

Joel Salatin
Summer 2007

The Past

Chapter 1

The Original Essay - Acres USA

In 2003 I was asked to be part of an international group to visit Laverstoke Farm in Great Britain, home of former Formula 1 racing champion Jody Scheckter, to discuss and forecast the pressing needs of a heritage-based food system. American gardening icon Eliot Coleman recommended that I be included in the group, and I held onto his coattails the entire trip.

Dan Barber, owner-chef of internationally-acclaimed Blue Hill Restaurant flew over to add his culinary artistry to the storybook pastoral setting an hour's drive from London. Jody's idea was to assemble, in his words, "the best of the best" farmers from around the world to identify where the clean food movement had been, where it was then, and where it needed to go. Kind of a visioning forum.

Taking our meals around Jody's magnificent 30-seat dining room table, we enjoyed the remarkable setting, a courteous and lavish host, and deep food-system conversations. On the final day, in the final session, we wrestled with the question, "What is the biggest impediment facing your farm specifically and the nonindustrial food system generally?"

We went systematically around the table as each participant critiqued current weak links. Several farmers mentioned labor; some mentioned cheap organic global trafficking; others mentioned consolidation and regulation in seed companies. As my turn approached, I was desperate for a way to succinctly capture how I wrestled with the regulatory and bureaucratic hurdles that constantly swarmed over us. When my turn came, I just blurted out, "Everything I want to do is illegal. "

Everyone at the table burst into laughter. They figured I wanted to smoke dope or run off with the neighbor's wife. I was actually taken aback by their response, and realized how funny it would sound to the average person. After the laughs and good-natured verbal jabs subsided, I said, "I'm serious about this." And what poured out of my soul was the accumulated frustration I had experienced with bureaucrats. Soon, every head at the table nodded in agreement and with understanding. Indeed, it was almost an epiphany for all of us as we realized this common denominator among industrialized nations – the so-called developed countries – that universally stifles what people in non-developed countries enjoy as a staple of their diet: local, indigenous, home-processed foods.

Upon returning home, I resolved to write an essay for ACRES USA magazine using the title I had blurted out in that room. It ran in the September issue of 2003, and has been by far the most quoted and most reprinted essay I have ever written. Chapter One of this book, fittingly, should be an unedited version of that original essay that always seems to bring a smile and nods of assent from anyone trying to create local food systems.

"EVERYTHING I WANT TO DO IS ILLEGAL"

Everything I want to do is illegal. As if a highly bureaucratic regulatory system was not already in place, 9/11/01 fueled renewed acceleration to eliminate freedom from the countryside. Every time a letter arrives in the mail from a federal or state agriculture department, my heart jumps like I just got sent to the principal's office.

And it doesn't stop with agriculture bureaucrats. It includes all sorts of government agencies, from zoning, to taxing, to food inspectors. These agencies are the ultimate extension of a disconnected, Greco-Roman, western, egocentric, compartmentalized, reductionist, fragmented, linear thought process.

1. ON-FARM PROCESSING. I want to dress my beef and pork on the farm where I've coddled and raised it. But zoning laws prohibit slaughterhouses on agricultural land. For crying out loud, what makes more holistic sense than to put abbattoirs where the animals are? But no, in the wisdom of western disconnected thinking, abbattoirs are massive centralized facilities visited daily by a steady stream of tractor trailers and illegal alien workers.

But what about dressing a couple of animals a year in the backyard? Why is that a Con-Agra or Tyson facility? In the eyes of the government, the two are one and the same. Every T-bone steak has to be wrapped in a half-million dollar facility so that it can be sold to your neighbor. The fact that I can do it on my own farm more cleanly, more responsibly, more humanely, more efficiently, and more environmentally doesn't matter to the government agents who walk around with big badges on their jackets and wheelbarrow-sized regulations tucked under their arms.

Okay, so I take my animals and load them onto a trailer for the first time in their life to send them up the already clogged interstate to the abattoir to await their appointed hour with a shed full of animals of dubious extraction. They are dressed by people wearing long coats with deep pockets with whom I cannot even communicate. The carcasses hang in a cooler alongside others that were not similarly cared for in life. After the animals are processed, I return to the facility hoping to retrieve my meat.

And when I return home to sell these delectable packages, the county zoning ordinance says this is a manufactured product because it exited the farm and was re-imported as a value-added product, thereby throwing our farm into the Wal-Mart category, another prohibition in agricultural areas. Just so you understand this, remember that an abattoir was illegal, so I took the animals to a legal abattoir, but now the selling of said products in an on-farm store is illegal.

Our whole culture suffers from an industrial food system which has made every part disconnected from the rest. Smelly and dirty farms are supposed to be in one place, away from people, who snuggle smugly in their cul-de-sacs and have not a clue about the out-of-sight-out-of mind atrocities being committed to their dinner before it arrives in microwavable four-color labeled plastic packaging. Industrial abbattoirs need to be located in a not-in-my-backyard place to sequester noxious odors and sights. Finally, the retail store must be located in a commercial district surrounded by lots of pavement, handicapped access, public toilets and whatever else must be required to get food to people.

The notion that animals can be raised, processed, packaged and sold in a model that offends neither our eyes nor noses cannot even register on the average bureaucrat's radar screen. Or, more importantly, on the radar of the average consumer advocacy organization. Besides, all these single-use megalithic structures are good for the gross domestic product. Anything else is illegal.

2. ON-FARM SEMINARS AND AGRI-TAINMENT.

In the disconnected mind of modern America, a farm is a production unit for commodities; nothing more and nothing less. Because our land is zoned agricultural, we cannot charge school kids for a tour of the farm because that puts us in the category of "Theme Park." Anyone paying for info-tainment creates "Farmadisney," a strict no-no in agricultural zones.

Farms are not supposed to be places of enjoyment or learning. They are commodity production units dotting the landscape like factories are manufacturing units and office complexes are service units. In the government's mind, integrating farm production with recreation and meaningful education creates a warped sense of agriculture.

The very notion of encouraging people to visit farms is blasphemous to an official credo that views even sparrows, starlings, and flies as a disease threat to immuno-compromised plants and animals. Visitors entering USDA-blessed production unit farms must run through a gauntlet of toxic sanitation dips and don moonsuits in order to keep their germs to themselves. Indeed, people are viewed as hazardous foreign bodies at Concentrated Animal Feeding Operations (CAFOs).

Farmers who actually encourage folks to come to their farms threaten the health and welfare of their fecal concentration camp production unit neighbors, and therefore must be prohibited from bringing these invasive germ-dispensing humans onto their landscape. In the industrial agribusiness paradigm, farms must be protected from people, not to mention free range poultry.

The notion that animals and plants can be raised in a way that their enhanced immune system protects them from kindergartners' germs, and that the animals actually thrive when marinated in human attention, never enters the minds of government officials dedicated to protecting precarious production units.

3. COLLABORATIVE MARKETING. I have several neighbors who produce high quality food or crafts that complement our own meat and poultry. Dried flower arrangements from one artisan, pickles from anther, wine from another, and first class vegetables from another. These are just for starters.

Our community is blessed with all sorts of creative artisans who offer products that we would love to stock in our on-farm retail venue. Doesn't it make sense to encourage these customers driving out from the city to be able to go to one farm to do their rural browsing/purchasing rather than drive all over the countryside? Furthermore, many of these artisans have neither the desire nor time to deal with patrons one-on-one. A collaborative venue is the most win-win, reasonable idea imaginable--except to government agents.

As soon as our farm offers one item not produced here, we have become a Wal-Mart. Period. That means a business license, which is basically another layer of taxes on our gross sales. The business license requires a commercial entrance, which on our country road is almost impossible to acquire due to sight distance requirements and width regulations. Of course, zoning prohibits businesses in our agricultural zones. Remember, people are supposed to be kept away from agricultural areas--people bring diseases.

Now, if we could comply with all the above requirements, a retail outlet carries with it a host of additional regulations. We must provide designated handicapped parking, government-approved toilet facilities (our four household bathrooms in the two homes located 50 feet away from the retail building do not count)--and it can't be a composting toilet. We must offer X-number of parking spaces. Folks, it just goes on and on, ad nauseum. All to just help a neighbor sell her potatoes or extra pumpkins at Thanksgiving. I thought this was the home of the free. In most countries of the world, anyone can sell any of this stuff anywhere and the hungering hordes are glad to get it. But in the great US of A we're too sophisticated to allow such bioregional commerce.

4. EMPLOYING LOCAL YOUNGSTERS AND INTERNS. Any power tool--including a cordless screwdriver--cannot be operated by people under the age of 18. We have lots of requests from folks wanting to come as interns, but what do we call them? The government has no category for interns or neighbor young people who just want to learn and help out.

We'd love to employ all the neighbor young people. To our child-fawning and worshiping culture, the only appropriate child activity is recreation, sitting in a desk, or watching TV. That's it. That's the extent of what children are good for. Anything else is abusive and risky.

And then we wonder why these kids grow up bored with life and nothing to do. Our local newspaper is full of articles and letters to the editor lamenting the lack of things for young people to do. Let me suggest a few things: dig postholes and build a fence, weed the garden, plant some tomatoes, split some wood, feed the chickens, wash eggs, prune the grapevines, milk the cow, build a compost pile, grow some earthworms.

These are all things that would be wonderfully meaningful work-experience for the youth of our community, but you can't just employ people anymore. A host of government regulatory paperwork surrounds every "could you come over and help us...?" By the time the employer complies with every Occupational Safety and Health Administration requirement, posts every government bulletin requirement, withholds for taxes, shoulders

Unemployment Compensation burdens, medical, Child Safety regulations–he can't hire anybody legally or profitably.

The government has no pigeonhole for this: "Hi, I'm a 17-year-old homeschooler and I want to learn how to farm. Could I come and have you mentor me for a year?" What is this relationship? A student? An employee? If I pay a stipend, the government says he's an employee. If I don't pay, the Fair Labor Standards board says it's slavery, which is illegal. Doesn't matter that the young person is here of his own volition and is happy to live in a tee-pee. Housing must be permitted and up to code. Enough already. What happened to the home of the free?

5. BUILD A HOUSE THE WAY I WANT TO. You would think that if I cut the trees, mill the logs into lumber, and build the house on my own farm, I could make it however I wanted to. Think again. It's illegal to build a house less than 900 square feet. Period. Doesn't matter if I'm a hermit or the father of 20. The government agents have decreed, in their egocentric wisdom, that no human can live in anything less than 900 square feet.

Our son got married last year and wanted to build a small cottage on the farm, which he now oversees for the most part. Our new saying is "He runs the farm, and I just run around." The plan was to do what Mom and Dad did for Teresa and me--trade houses when children come. That way our empty nest downsizes and the young people can upsize in the main family farmhouse. Sounds reasonable and environmentally sensitive to me. But no, his little honeymoon cottage--or retirement shack--had to be a 900 square foot Taj-Mahal. A state-of-the-art accredited composting toilet to get away from a septic system and sewer leach field was denied.

And when the hillside leach field would not meet agronomic standards and we had to install it in the flood plain, I asked the health department bureaucrat why. He said that essentially the only approvable leach fields now are alongside creeks and streams because they are the only sites that offer dark enough colored soils. Sounds like real environmental stewardship, doesn't it?

Look, if I want to build a yurt of rabbit skins and go to the bathroom in a compost pile, why is it any of the government's business? Bureaucrats bend over backwards to accredit, tax credit, and offer money to people wanting to build pig city-factories or bigger airports. But let a guy go to his woods, cut down some trees, and build himself a home, and a plethora of regulatory tyrants descend on the project to complicate, obfuscate, irritate, frustrate, and virtually terminate. I think it's time to eradicate some of these laws and the piranhas who administer them.

6. OPTING OUT OF THE SYSTEM. I don't ask for a dime of government money. I don't ask for government accreditation. I don't want to register my animals with a global positioning tattoo. I don't want to tell officials the names of my constituents. And I sure as dickens don't intend to hand over my firearms. I can't even use the "O" word (Organic).

On every side, our paternalistic culture is tightening the noose around those of us who just want to opt out of the system. And it is the freedom to opt out that differentiates tyrannical and free societies. How a culture deals with its misfits reveals its strength. The stronger a culture, the less it fears the radical fringe. The more paranoid and precarious a culture, the less tolerance it offers.

When faith in our freedom gives way to fear of our freedom, silencing the minority view becomes the operative protocol. The native Americans silenced after Little Big Horn simply wanted to worship in their beloved Black Hills, use traditional medicinal herbs to cure diseases, educate their children in the ways of their ancestors, and live in portable homes rather than log cabins. By that time these people represented absolutely no threat to the continued Westernization and domination of the North American continent by people who educated, vocated, medicated, worshiped, and habitated differently.

But coexistence was out of the question. Just like the forces that succeeded in making it illegal for me to use the "O" word, the western success at Wounded Knee quashed the little guy. What does the Organic Trade Association have to fear from me using the "O" word? If society really wants government certification, my little market share will continue to deteriorate into oblivion. If, however, the certification effort represents a

9

same-old, same-old powergrab by the elitists to exterminate the fringe players, it is merely another example of fear replacing faith.

Faith in what? Faith in diversity. Faith in each other. Faith in people's ability to self-educate, thereby making informed decisions. Faith in seekers to find answers. Faith in marketplace dynamics to reward integrity and not cheating. Faith in creation to heal. Faith in healthy plants and animals to withstand epizootics. Faith in earthworms to increase fertility. Faith in communities to function efficiently and honorably without centralized beltway interference. Faith in ACRES USA to come every month with a cornucopia of insight and information.

Our culture's current fear of bioterrorism shows the glaring weakness of a centralized, immuno-deficient food system. This weakness leads to fear. Demanding from on high that we irradiate all food, register every cow with government agencies, and hire more inspectors does not show strength. It shows fear. It does not show faith in the natural balance between good bugs and bad bugs when plants and animals are raised in habitats that allow them to achieve their individual physiological expression. It does not show faith in relationship marketing, farm-gate sales, or Community Supported Agriculture to build protection and integrity into the food system.

Indeed, official policy views all these minority production and marketing systems that have been shown faithful over the centuries to be instead something that threatens everyone and everything. As a tee-pee loving, herb healing, home educating, people-loving, compost-building, retail farmer, I represent the real answers but real answers must be eradicated by those who seek to build their power and fortunes on a lie. The lie being that genetic integrity can be maintained when corporate scientists begin splicing DNA. The lie that, as Charlie Walters says, toxic rescue chemistry is better than a balanced biological bath. The lie that farms are disease-prone, unfriendly, inhumane places and should be zoned away from people.

Those of us who desire to opt out--both patrons and producers--must pray for enough cleverness to circumvent the system until the system cannot sustain itself. Cycles happen. And because things are this way today does not mean they will be this way next year. Hurrah for that.

Often the greatest escapes occur at the moment the noose becomes tightest. I'm feeling the rope, and it's not very loose. Society seems bound and determined to hang me for everything I want to do. But there's power in truth. And for sure, surprises are in store that may make society shake its collective head and begin to question some seemingly unalterable doctrines. Doctrines like the righteousness of the bureaucrat. The sanctity of government research. The protection of the Food Safety and Inspection Service. The helpfulness of the USDA.

And when that day comes, you and I can graciously offer our society honest food, honest ecology, honest stewardship. May the day come quickly.

Chapter 2

Raw Milk and Dairy

When I was a late teen and contemplating how to become a full-time farmer on our family's farm, I remember as if it were yesterday that milking ten cows and selling the milk at regular supermarket prices would be all I needed.

We always milked a couple of Guernsey cows by hand. Few things are as relaxing as the cow's quiet breathing and cud-chewing accompany the squirt-squirt of the milk sudsing into the bucket. Often a cat sat nearby, ready to catch a target-practice squirt aimed squarely into its mouth.

We brought the milk into the house, strained it through a paper filter to catch any hair or dirt pieces, and refrigerated it in big pots to let the cream rise. Using a shallow dipper, we skimmed off the cream and used that for butter, whipping cream, and ice cream. We made cottage cheese and yogurt with what we couldn't drink. Sometimes we had so much milk we fed it to a couple of pigs.

We sold butter and cottage cheese to neighbors and at the Curb Market. A precursor of today's Farmers' Markets, the Staunton Curb market began during the depression era as a means for cash-deprived farmers to earn money. Farmers never lacked for food, but they did lack for cash.

As food safety laws proliferated during the next few decades, the Curb Market enjoyed wonderful exemptions because it was under the authority of the Extension Service. In order to sell there, vendors had to join an Extension Homemaker's Club—originally called the Home Demonstration Clubs, until the word demonstration became a liability during the Vietnam war protest era—and men had to be involved in Extension programs. A memorandum of agreement between the food police and Extension created a congenial arrangement between the two organizations.

The food police were willing to exempt the Curb Market vendors from the infrastructure requirements because the vendors were getting the latest greatest government food safety thinking through the Extension Service. By the time I came on the scene as a 14-year-old in 1971, the once-bustling Curb Market had dwindled to two elderly matrons and was located in the old train warehouse district of town. During the 1950s and 1960s, the market dwindled as the culture became enamored of everything industrial. The allure of the supermarket, boughten bread (remember that one from the Little House on the Prairie Books?), and white sugar tempted people to abandon their roots and throw common sense out the window. Even breast feeding babies, that contemptibly barbaric act, gave way to enlightened Infamil and Similac.

Consumers summarily pulled up their food anchors and sailed for bright lights and citified facades. During the late 1960s and especially the early 1970s, of course, this trend reversed itself in the hippie movement yearning for meaning, soul, and Woodstock. But during the 20-30-year cultural love affair with everything unnatural, all of this magnificent community-based food infrastructure was lost. And the ensuing problems created by this pseudo-food alternative foisted a mountain of bureaucracy and regulation on everything involving food. That is where we are today.

Opening each Saturday morning at 6 a.m., the Curb Market at one time sported some 40 vendors and was the retail action place in the community.

I joined 4-H in order to qualify for the exemptions, and during the 1970s, my high school years, our family sold butter, cottage cheese, yogurt, uninspected fresh beef and pork that we either dressed and butchered at home or at a custom facility, and eggs. I would dress my spent laying hens at home, cook them in the oven in a big roaster pan, pull off the meat, cut it up into quart freezer containers, and sell it as pre-cooked boneless chicken. We sold vegetables and home-processed rabbit.

One of the two ladies sold potato salad, pound cake, bread, cakes, and pies. The other lady sold some baked goods but also home-processed and cured pork and lots of vegetables. This was the Viet Nam era, just before the hippies traded their pot and free love for organic food and marriage. But even then, the Curb Market was a toehold for local and artisanal fare, and attracted a steady clientele. It was an anachronism when I was there. The fledgling poultry industry was just getting started; hence the product was not yet bad enough to drive away masses of consumers like it does today. Food borne bacteria hardly existed; nobody had heard of cloning, genetic engineering, irradiation, or mad cows. Avantis and Monsanto were high-risk start-up companies.

When I went to college, we closed down our booth. During that four years, the other two ladies quit and the Curb Market was no more. Tragic. Tragic. I often wonder what would have happened had I not gone to college, but rather stayed at the curb market to enjoy grandfather status as the food laws became worse and worse. I'm confident that my presence there gave the two grandmas renewed enthusiasm for their business and perhaps kept the market going a little longer than it would have otherwise. By a long, long shot, I am now the last surviving vendor of the Staunton Curb Market, a distinction I wear with honor and pride. Those were the glory days.

That is where our family learned the power of value adding. We did not price things much above supermarket rates, but by selling retail, we made a nice gross margin even as small farmers. That was the spark that has carried our family through. Of course, at that time the American family still ate many meals around a dining room table. Chicken franks and boneless, skinless breasts had not arrived on the supermarket shelves yet. Home cooking was still home cooking, not just micro-waving heat-and-eat packages.

So here I was, late teen, dreaming of farming, trying to figure out how to make it happen. Mom and Dad worked off the farm, he as an accountant and she has a high school health and physical education teacher, to pay for the land. The mortgage had been paid. But how could I get cash? I didn't need much-- $10,000 was plenty. I had neighbors, old-timers they are called, who milked 15 cows by hand. All of them had forearms the size of vegetarians' thighs.

I knew I could milk 10 cows by hand, earn $1,000 per cow, and make a fine living. Only one little problem: it was illegal. If there is one product that has been demonized in our modern day, it is raw milk. All sorts of diseases allegedly trace their origins to this hazardous product.

Milk-related illnesses, though, are a blip on the screen of human history. They did not exist before the industrial revolution. Here's the nutshell. Without refrigeration, and without rural electrification and refrigerated trucks, food systems were more local in, say, 1910. Breweries grew up in the cities to service the needs of the locale. These distilleries generated byproducts. In those early days of industrial farming, large dairies sprang up around the distilleries as a garbage disposal for their byproducts called distiller's grains.

These byproducts change the pH of the rumen to more acidic, which changes all sorts of balances within the cow. A confluence of factors created a pathogen-laden environment for humans. In the 19-teens, the biggest issue confronting cities was horse manure. In fact, many planners were prophesying that cities would literally be flooded with manure. No one at that time envisioned that it would cease to be a problem within one short decade with the advent of the automobile.

Another factor complicating things was the return of thousands of young men from foreign countries carrying exotic diseases—like Spanish Flu—in the aftermath of World War I. All

15

of this manure, disease, lack of human hygiene, and sick cows combined to create an unprecedented pathogen proliferation. Tuberculosis, undulant fever, whooping cough, and scarlet fever are all results of this deadly confluence.

Interestingly, a new set of diseases is now threatening our population as a result of industrial food. These are primarily obesity, food borne pathogens like *E.coli, camphylobactyer, lysteria, bovine spongiform encephalopathy,* and *salmonella,* along with Type II diabetes, heart disease, and cancer. These things were unheard of before 1900, when industrial food began its steady encroachment on the world food system.

In fact, in his blockbuster book <u>Guns, Germs, and Steel</u>, Jared Diamond points out that the close connection between large domestic animals and people in Medieval Europe created pathogenic immunities in people. I wonder if we aren't entering a new phase, where our culture is becoming disconnected from the physical to such an extent that we are losing immunological ground. Intuitively, we can appreciate the need for callouses, blisters, splinters, cuts, and bruises, not to mention eating some dirt on carrots or ingesting small amounts of impurities in milk or cider to keep our immune systems exercised.

That we've become an exercise-machine-only culture, not to mention Nintendo and cyber-creation, is probably setting us up for a collision course with newly virulent industrial-strain pathogens at the very time when we are least able to handle them. We should be rolling in the dirt, gardening, wrestling with some brambles and skinning animals for supper. These are important immune system builders.

As a culture, then, we went through a spasm of adjustment to these new pressures, worked through them, and emerged on the other side. But during this thirty-year adjustment (roughly 1915-1945), we created a body of perceptions about milk; namely, that it was not inherently wholesome when it comes out of the cow. Secondly, rather than stop feeding cows inappropriate waste products, government food safety officials imposed pasteurization rules that mandated a whole new set of infrastructure between teat and drinking glass.

As usual, rather than cleaning up our collective act, we as a culture went blindly forward with a paradigm that disrespected and dishonored nature but assumed we could technologically fix the problem such dishonor engendered. What an egocentric view. Such arrogance will naturally be attacked by bugs and bacteria to balance out the egocentrics.

As I gradually began to realize just how plausible it would be to hand milk ten cows and sell their milk at retail to my neighbors, I first became frustrated and then angry that such an option was illegal. Here again, our culture had criminalized the righteous act while requiring the evil act—mandatory pasteurization. If you want to find out more about the raw vs. pasteurized debate, plenty of material exists on that issue without me recapping it here. What I want to convey is the dream-dashing this criminalization creates.

Never mind that raw grass-fed milk is as safe and wholesome as it has been throughout human history. Never mind that customers by the thousands yearn for this unadulterated product. In fact, never mind that half the states allow its sale to some degree. It's just illegal to sell in Virginia, period.

Even if we were to move forward with cheese or some milk product, we would still need a license and inspected facility. A friend who ran a Grade A dairy wanted to make cheese. But by the time he installed all the required machinery and hardware, it would have cost them $100,000 to make one pound of cheese. End of dream. He continues to struggle, barely making ends meet. I'd love to buy his cheese, even if he made it in the kitchen sink. And that's important to understand.

Some might ask, "Why don't you just put in the infrastructure and comply with the requirements? Why do you have to be such a stick-in-the-mud about all this? If you're going to run a business, then just install what a business should have and quit refusing to comply. What's wrong with a little investment if you want to operate a business? You just want to have all the benefits of the business without making the requisite investment. You want a shortcut. Phooey on you."

Here's the answer, and it deals with the whole issue of innovation. All new things start small. Mighty oak trees begin from a tiny acorn, not 20-foot baby trees. Humans are born as babies, not teenagers. Innovation demands a prototype first, and a prototype must be as small as possible.

How do I know if I have a cheese that people will want unless I can experiment with a few pounds and try to sell some to folks? How do I know I have a decent ice cream until I make some and sell it to taste testers? Innovation demands embryonic births. The problem is that complying with all these codes requires that even the prototype must be too big to be birthed. In reality, then, what we have are still-birth dreams because the mandated accoutrements are too big.

Here's an analogy. We're all familiar with eBay. A true American cultural phenomenon, and who would want to see it disappear? Just imagine the following requirements before you could list an item on eBay:

- License certifying you are qualified to operate your computer.
- Occupational Safety and Health Administration (OSHA) certificate certifying that your computer desk and office are splinter-free, handicapped accessible, and non-injurious to you.
- Electrician certificate verifying that the electrical cords running to the computer are up to code.
- Fire marshal license that you have a fire extinguisher in working order, properly checked, on the wall in case your hot item becomes too hot to handle.
- Government labeling verification that your description of said item is indeed accurate and not misrepresented in any way.
- Building inspection license indicating the structural soundness of the room and computer desk in the immediate vicinity of where you will be doing eBay business on your computer. What if the room caved in on you right when a hot buyer was on the line? The potential buyer might

suffer emotional trauma in the fallout of your room falling in.

• Securities and exchange commission license for your banking and accounting showing that you can indeed transact financial matters and that you are properly indemnified against product liability.

I could go on in this vein for awhile, but I think the point is abundantly clear—would eBay have ever existed, had all these requirements been placed on it? Indeed, would anyone today use it were these requirements in place? And yet it is an entirely self-policing system for all of these elements. How can that be? The reason is that it is transparent and has a dramatic feedback loop for customer comments. A shyster simply doesn't last long in this market climate.

eBay works because anyone can try it without investing in any extra infrastructure. It utilizes already-owned computers. It's a true free market entity, and of course drives retailers crazy. The retail industry lobbies hard to clamp down on this freewheeling phenom, but fortunately eBay freedom still exists. Because anyone with a computer and an item can launch a transaction, it epitomizes the embryonic prototype. And its dramatic success shows the validity of this concept.

The same is true with any business. Only the foolish or independently wealthy can afford to birth a 100-pound baby. Most of us need some time to grow into our adolescence. And that means we must preserve the freedom to access the market on a tiny scale or innovation cannot occur.

Today's big food businesses began from the tailgate of a pickup truck in the 1940s and 1950s before food safety laws became as onerous as they are today. If we do not preserve this type of market access for future generations, we not only keep a generation of young people from realizing their dreams, but we deny the entire culture the innovative prototypes necessary to truly solve problems. If every baby must be born too big to be viable, the babies die. And it's the babies that have the new ideas, that dare to question old paradigms, that move a culture forward.

And to me, one of the benchmark attitudes that distinguishes evil hearted people from good hearted people is their attitude toward the babies. In the natural progression of business, a prototype outgrows the kitchen sinks, for example, and eventually does require a separate building and some snazzy stainless steel furniture. For example, on our farm, we did not install a walk-in freezer and cooler until we had 13 home-style chest and upright freezers. Then we made the leap to the upgrade, and a huge investment it was. But we grew into it.

If my home-kitchen pound cake prototypes are well received, eventually I outgrow the kitchen and build a bigger one, with bigger mixing bowls and larger sinks. But if I must build a $50,000 freestanding commercial kitchen before I even experiment to find out if baking pound cakes is something I enjoy, I will never bake the first one. And herein lies the great conundrum of this book. My passion is to re-create in our culture a sense of loss. I lie awake at night trying to figure out how to get people to understand what we've lost.

I'm desperate for a food NRA (National Rifle Association). The reason the NRA exists and is such a powerful organization is because such a huge number of Americans own guns. And they don't want to lose them. The fear of the loss motivates us to join the NRA and to passionately defend "America's First Freedom."

The only reason we don't have a food NRA is because it's hard to create a need for something that does not exist. It's hard to get people passionately exercised about something they don't have. In a nutshell, the goal of this book is to give Americans an insatiable appetite for something they don't have. I want folks to leave these pages angry that they've been denied something righteous, something healthful. I want folks incensed that their government has sold our collective freedom birthright for a bowl of global corporate outsourced pottage.

If you've never tasted real homemade jams and jellies, you don't know what you're missing. If you've never tasted an on-farm processed pastured broiler, then Whole Foods organic will do just fine, thank you. But once you've tasted the real deal, nothing else satisfies. I don't know how many times our customers tell us, "You've ruined us," meaning that they just can't go back to normal food, even certified empire organic. I had a restaurant chef a week ago get all weak-kneed on me when he found out that we were out of eggs. "What am I supposed to do?" he lamented.

Eye hath not seen nor ear heard, to paraphrase a Biblical description of God's plans, what could be if local food entrepreneurs were freed up to access their neighborhoods with homemade, artisanal food. Let me just for a moment describe what would happen.

We'd enjoy muffins from our neighbor's kitchen, pickles, salsa, and baked goods. We'd have locally-grown and cured ham and bacon. Local beef jerky would accompany our kiddos' lunch boxes. Frozen heat-and-eat quiche made from overabundant pastured spring eggs from the neighbor's flock would offer quickie meals on soccer night. Chicken pot pies, made with grandma's recipe in our neighbor's kitchen could be purchased during that mad-dash-home-from-work-what's-for-supper panic. We could enjoy a Delmonico steak from a pastured steer that never stepped onto a trailer to be co-mingled at a slaughterhouse with animals of dubious extraction while awaiting slaughter, but rather was killed on its home farm in reverence by the farmer who cared for it. Tender and beyond description.

All of this washed down with wine from the neighbor's grapes, fermented lovingly in his basement. Under the watchful eye of the children and the family cat. Cheese, all sorts of fresh and aged, straight from the neighbor's ten-cow pastured dairy herd. And ice cream to die for, from heavy cream and molasses grown, milled, and canned two miles down the road. Are you salivating yet?

You see, an imbedded local food system could actually exist in the midst of subdivisions and strip malls. Wherever a few unpaved square feet poked through, edible plants and animals could be grown and processed for the neighborhood. And without the expensive labeling, packaging, and processing infrastructure requirements, this food could be sold at regular supermarket prices, and it would be infinitely better. Virtually all of the processed foods currently sold at the supermarket could be supplanted with community-based entrepreneurial fare. Does your heart ache for this? Mine does.

Well, why can't we just have this? Because everyone is paranoid of the unscrupulous. I was eating dinner with a couple of Sustainable Agriculture professors from the University of Iowa and they were appalled that I would suggest we allow unfettered market access to cottage-based food producers. "But, but, but, you can't just let food on the market that isn't regulated," they spluttered. "What about the dirty kitchens? The dishonest people?"

It was a lively discussion, and we concluded with my question, "Okay, then what would you do? Sleep on it and tell me in the morning."

We parted amicably—my goodness, these were my friends. I realized that if I couldn't sell them the idea of an unfettered entrepreneurial food system, then I certainly couldn't sell the idea to anyone else. These folks knew all about the evils of industrial food systems, the environmental footprint of outsourced food, and the failure of U.S. agriculture policy. I was frankly shocked that they, too, had swallowed this paranoia that assumes everyone selling an item is untrustworthy. So I put the question to them and they promised to have an answer for me by morning on our shuttle ride to the airport.

Morning came. "Okay, what did you come up with?"

"Nothing. We don't have an answer." That was it. Here these people are nationally-recognized, published professors of sustainable agriculture, and they have no idea how to create a local food system. Well-versed in low-chemical production models, companion planting techniques, and composting, they failed to make the connection between stewarding the land and stewarding the farmer's profitability; and, ultimately, the community's sustainability via a local food web.

That conversation helped me to understand the dilemma we face as people like me preach local food systems. The current alternative food movement grew out of the back-to-the-land Mother Earth movement of communes, hippies, and Woodstock. Completely disenfranchised from Wall Street and the establishment, these visionaries sought a different direction for our country through government.

We can cure pollution with an Environmental Protection Agency. We can cure abusive animal practices with more regulations. We could heal the forests by taking land away from the private sector and expanding wilderness areas. We could end poverty with more welfare and social programs. It was virtually a top-down mentality. Indeed, this is still evidenced by the organic movement, that asked for government certification. People like me prophesied that when the government controlled the movement, the little guys would be squeezed out, the standards would gradually be adulterated until organic meant nothing, and it would simply be a way for multinational globalists to hijack organics.

And at the time, people like me were laughed to scorn by those who really believed the government could help out. A few tax dollars here and there would be wonderful. This idea is still alive and well. I was recently contacted by a leading sustainable agriculture lobbying group to sign onto a letter delineating policy positions for the new farm bill. The first part demanded a cap on subsidies to individual farms. Many people don't realize that wealthy, wealthy land owners and mega-corporate outfits enjoy the lion's share of farm subsidies. I was okay with caps.

But the second part of the lobbyists' agenda demanded that the monies saved from capping would be given to sustainable farmers instead. What? How disingenuous can we be? Here our side has been decrying this inappropriate subsidy for decades, and then we demand to pig out at the public trough too? Where's the high moral ground in that position? To the proponents of an industrial food system, if we lobby to partake we are no better than they.

Anyway, the alternate food system grew out of this "government-can-fix-it" mentality. For lack of better terms and just to push this discussion to a conclusion, I'll call this a more liberal idea as opposed to the conservatives, who take a more limited government position.

The leaders of the clean food movement, by and large, still adhere to this liberal idea and vote Democratic. Many have voted for Ralph Nader and the Green Party. The problem with the government fix is that while it may start sincerely, by the time it gets implemented on the ground and has been through the sieve of corporate dinners, it hurts the little guys and helps the big guys.

Case in point: consumer advocates proposed several years ago to require daily testing for pathogens on poultry. Each slaughter facility would have to take a swab test. One catch: it would cost $300. Now to Tyson or Pilgrim's Pride, that's a spit in the ocean. Their electric bill to maintain cool drinking water is more than $300 a day. But to a small pastured poultry outfit like ours, that's the whole day's profit. End of the little guy. That is reality, and unless you've had a small business, it's hard to appreciate the unintended consequences of these wonderful-sounding proposals.

Of course, the other side is equally discriminatory. When the Republicans propose business incentives and tax breaks for business, they end up lining the pockets of the big players. And it's such a common occurrence now that "Corporate Welfare" is synonymous with Republicanism. And it's rotten.

As an example, I was interviewed by Pat Buchanan on his radio show when Bill Clinton was in the White House. It was shortly after news hit the food sections that Bill and Hillary hired a French chef who vowed that he would serve only "free range chicken." That was just too juicy for the conservatives to let go. Buchanan found me and put me on the show. He asked me what the difference was between regular chicken and free range—of course, I didn't use free range, I used pastured.

I explained that for one thing, our chickens didn't do drugs. "Well, why do they give drugs in the big houses?" he asked.

"For one thing, it makes them grow faster..."

He cut me off: "How could anything be bad that makes them grow faster?"

I was incredulous. Cancer is fast growth. But this shows the conservative disrespect for biological and environmental realities. For crying out loud, we should be able to all understand that if growing it faster, bigger, fatter, cheaper were a good goal, we'd all aspire to be the fattest person in the room. But this is the problem with the conservatives. They measure growth only in terms of Gross Domestic Product. Truly, if the liberals have no head, the conservatives have no heart. Okay, I have the rare joy of being able to make everyone mad at the same time. Rare gift, I know.

To just assume that the faster we can grow a chicken, the better, is ridiculous. All paradigms exceed their efficiency. Single-trait selection always creates problems in other areas. What we want is balance. But this is the mentality of America's business climate: faster growth is better.

Back to the original question for the liberals and most people in America: How do we punch through this skepticism that assumes that if I sell you something, I'm trying to rip you off? That's the mainline mentality. And it's the mentality that is keeping you and me from enjoying a plethora of community-based eats and treats. I am encouraged by a new awareness within the more liberal political spectrum that maybe the government doesn't have all the answers.

And I'm equally encouraged by an awareness among conservatives that maybe taking care of the earth is as important as taking care of chief financial officers. These are positive developments indeed, and give me reason to hope that maybe we can punch through this dilemma.

The problem is that as soon as we posit "the government is responsible" we open Pandora's box on a host of regulations that require a host of bureaucrats and deny a host of dreamers from fulfilling their dreams.

Ultimately, this issue is about how we view the fringe, or the nonconformists. If it's the government's responsibility to make sure that no person can ingest a morsel of unsafe food, then only government-decreed food will be edible. And when that happens, freedom of choice is long gone, because the credentialed food will be what the fat cats who wine and dine politicians say that it is. In the name of offering only credentialed safe food, we will only be able to eat irradiated, genetically adulterated, inhumane, taste-enhanced, nutrient-deficient, emulsified, reconstituted pseudo-food from Archer Daniels Midland, "supermarket to the world."

And we will forget all about what could have been. The ten cows, their raw milk, butter, and ice cream will be long buried in the annals of American heritage, relegated to dusty pages of history books and woodcuts of a bygone era.

Chapter 3

PL 90-492

"**Y**our chicken processing is unsanitary and adulterated. We have to shut you down," said the chicken policeman. Actually, he was head of meat and poultry inspection for the Virginia Department of Agriculture and Consumer Services.

We had been processing and selling chickens for a couple of decades and we had an exemption number, Va. # 1001 (the first one ever given in the state) that allowed us to process up to 20,000 head of poultry per year in our non-inspected facility.

The federal exemption, known as Public Law 90-492, delineates what is known as the Producer-Grower Exemption. This federal law allows a producer-grower to process up to 20,000 head of poultry annually without government inspection. The only stipulation is that the processing be "sanitary and unadulterated." Obviously, those two words are completely objective. Everyone knows exactly what sanitary and unadulterated mean, I'm sure.

This exemption, written into the federal code during the major rewrite of the inspection laws in 1967, granted a numeric exemption for farmers to process their own birds and sell them. I have asked many insiders over the years how this wonderful exemption came to be, and the only answer is that at that time, the vertically integrated poultry industry did not yet exist and most poultry still came from small producers, many of whom processed the birds on the farm. Because it was politically impossible to close all of them down, a numeric concession enabled all these little operations to remain in business.

The same climate did not exist for cattle, hogs, sheep and goats, which had already gone through more marketing consolidation. And, certainly, fewer farmers processed their large animals. Rather, they used the multitudinous community abattoirs. Our county, at the time, had about two dozen neighborhood abattoirs for large animals, but these typically would not do poultry. As a result, many chicken farmers ended up processing their own birds. Since chickens and turkeys are not as heavy to pick up, they lend themselves to small-scale processing. Picking up a 1,000-pound steer to skin, gut and shove into a cooler is a big deal. To do that to a chicken is not a big deal.

As a side note, within just a couple of years after the 1967 rewrite, our county lost all but three of those neighborhood abattoirs. They did not go out of business because they suddenly became unprofitable. They went out of business because the new regulations were too onerous for the abattoirs to remain viable on their small volumes. We'll talk about this more in the beef chapter.

At any rate, the wonderful poultry exemption, which as of this writing is still the law of the land, created an opportunity for thousands of small poultry operations to remain in business and serve their communities. The beauty of this exemption was that it did not specify any infrastructure; it only specified that the processing be done in a sanitary manner and the product could not be adulterated. Adulteration is non-chicken things on the chicken—like dog and cat hairs or hay seeds. The other requirement is that the producer apply name and address on the label. In Virginia, the inspection service also mandated that the operator be given an exemption number so that each of these would be on file in the bowels of the state capital. That particular requirement has since been terminated because it is not part of the law.

A decade prior to the incident that put us on a collision course with the chicken police, our on-farm processing had been scrutinized by the former head of Virginia meat and poultry inspection, and he was duly impressed. So much so that after touring the farm's pastured poultry, looking at our little out-door processing shed, and hearing my vision for local food systems, he settled back on a sofa in our living room and said wistfully, "Joel, being here makes me want to be a young fellow again and stay on our family's farm and do it your way. Maybe I could have kept the farm, which is what I really wanted to do."

The amicable, reasonable interchange had a wonderful outcome. He quickly gave us our exemption number and a letter of commendation. Two years later he retired. And his replacement was a Pharaoh who knew not Joseph, if you get my drift. We were in a new dispensation. Now mind you, no laws had changed. Legally, everything was the same. The only thing different was a change of personnel.

Before we move forward in what ended up being a lengthy discussion, let me describe our processing shed. It's a concrete slab about the size of a two-car garage with six locust poles sticking up through the slab to support a simple metal shed roof. The slab slopes so all the water runs off into a 10,000 gallon lagoon. The lagoon has a simple shed roof over it as well so nothing can inadvertently fall in.

The furniture in the shed consists of a homemade rack that holds eight killing cones. The blood drains into a V-shaped trough that funnels it into a couple of buckets. A factory-made scalder-dunker loosens the feathers with hot water. Then a factory-made chicken picker with a rotating bottom and 3-inch long rubber fingers gets rid of the feathers. A white enamel sink-drainboard receives the plucked birds. During the picker-to-sink transfer, we cut off the feet and pull off the heads. This old sink was originally in our kitchen when our family came to the farm in 1961. When we modernized the kitchen in the 1970s we installed a double stainless steel sink, but like most scroungers, held onto the old piece "just in case."

Then the birds go onto a stainless steel evisceration table where we pull the innards out by hand before putting them onto the quality control table, which is our oldest piece of furniture. In about 1970, when I was only 13, I needed a chicken picker in order to process my spent hens from my 4-H laying flock. We put an ad in the rural electrification co-op magazine and got two hits. Hitching up our 1940s trailer to the 1963 Chrysler Newport with push-button console gearshift, we headed off to the Eastern Shore. Looking back on it now, I realize what a huge deal this was for Mom and Dad to empower their son with this chicken venture deal.

The Eastern Shore is a five-hour drive away. The man who had responded to the ad was a 75-year-old farmer who had raised chickens all his life and was getting out of it. He had an ancient galvanized table, scalder, picker, and bleeding cabinet with shackles, all for $75. We bought the whole kit and caboodle. The table and picker were treasures; the rest we discarded.

The table was simply a wooden table 2 feet wide and 6 feet long with galvanized sheet metal bent and soldered over it and a drain hole plumbed in the bottom. The picker was an old horizontal one-at-a-time variety, but it worked fine. We long ago outgrew the little picker, but we still use the table. It must be 70 years old. The galvanize long ago left, and it is now a beautiful antiqued rusty color, no doubt sealed forever with half a century of chicken oil. It is now our quality control (QC) table.

After the QC table, the birds go into a double-stainless steel sink from an old A&P that went out of business. Dad and I went to the auction and bought that in about 1980. From that initial chill tank, the birds go into a couple of large stainless steel chill tanks that a friend scavenged from a factory scrap metal pile for $5 apiece. That's the setup. And with eight people we can run 200 birds per hour, which happens to be just as efficient a chicken-per-person-hour as the big industrial processing facilities.

The open processing shed has several advantages:
- Direct sunlight's radiation is the most efficacious sanitizer.
- Walls would block sunlight and make lots of dingy, dark, damp corners to harbor pathogens that would then have to be sanitized with toxic cleaners.
- No artificial lighting needed. We don't process at night, so the open sides allow enough light in to enjoy natural lighting.
- We can see out. Surrounded by birds singing, the adjacent vegetable garden, the nearby pond usually sporting a couple of wild ducks or a Blue Heron, it makes an enjoyable place to work.
- Cool. Since our poultry processing is seasonal, we don't want to block off natural air currents with walls, and the shed is always cool, even on the hottest days.
- Simpler to build. When we built the shed, we simply didn't have the money to build a high-falootin' building. The construction was within my capabilities and we did it on a shoestring with poles cut from our woods and lumber scavenged from a barn demolition project at a neighbor's place.
- Conducive to education. Customers and wanna-bes can come and gather around to see the processing without cramping us inside. With walls, anyone who came to watch and learn, or get connected with their food, would be in the way.

The new chief disagreed with his predecessor: this facility was unacceptable. Of course, I immediately asked him the obvious: "Wait a minute, Bob. The law has not changed since we were issued a glowing bill of health a few years ago. What changed?"

"Me. I'm the new division head. "

"But you are administering the same law, right?"

"Yes. But I'm interpreting it differently. "

"Why now? Why today? What's the big deal?"

"I was just going through the files and found you and realized you probably were not in compliance." By the way, compliance is their favorite word. In any conversation with bureaucrats, they will say it about every fifth word.

"How so?"

"An open air facility is inherently unsanitary."

"So you never go on picnics? People have been processing meat and poultry in the open air for centuries. What's the big deal all of a sudden?"

"That's just the way I see it. And if one fly comes in and lands on one chicken, that is now an adulterated product."

"You mean if a fly lands on your hotdog at a picnic, you throw it away?"

"That's not the point. The point is what I think. And I think you don't comply with the sanitary and unadulterated standard. "

"So what would it take for me to comply?"

"You have to have walls."

"Why?"

"Because the regulations say so."

"Where?"

"Right here it says that doors and windows must be screened. Obviously, the assumption is that you have walls."

"I disagree. All it says is that if you have doors and windows, they need to be screened. That doesn't assume you have walls. Obviously, if you have walls, you have to get in somehow. And the regulation just says that those enclosures need to be screened. Now, I can go out and hang a window or a door on one of those poles, and put a screen in it if you'd like..."

For some reason, he didn't think this was funny. For the record, the rule he was citing was not in the exemption section, but was over a few sections in the "custom" slaughterhouse rules. That's when I'm processing some else's chickens. The rules there are far more stringent because now I'm not processing my own birds. The producer-grower exemption allows us to sell the birds to hotels, restaurants, institutions, and in Virginia, at least, to retail outlets as long as "EXEMPT PL 90-492" is written on the label, along with our name and address.

Also for the record, it's important to understand that these walls are not just any old walls. They must be impermeable so nothing can penetrate them and they can be washed off. Typically, that means expensive fiberglass material that comes in sheets like drywall or plywood, but is far more expensive. A guy in Ohio, unable to punch through this with the chicken police there, finally installed a hoophouse with plastic over his facility. The officials ranted and raved about it, but he just smiled and said, "It's not permeable." I love it when someone is creative enough to drive these dudes nuts.

Leaving the walls, I then asked him what next would be necessary for compliance. His answer, "Bathrooms."

"Why?"

"For your employees, and it's required."

"I don't have any employees. And we have four bathrooms within 50 feet—two at my house and two at Mom's house. And if I want to go #1, I just step around behind the tractor. What are you going to do about that?"

I guarantee you that if these people could outlaw taking a leak behind the shed, they sure would. It's gotten to where you can't even spit without a license.

"Well, what else is necessary?" I pressed.

"You need a changing room with 12 lockers."

"Why?"

"For compliance."

How do you carry on a conversation with these people? It just goes from ridiculous to asinine. I didn't bother asking him any more.

We finally arranged a showdown with his federal boss out of Philadelphia, our attorney, and the legislative aide for our state delegate. Our state senator had already made some phone calls, but could not attend the meeting.

When they all arrived, we ushered everyone into the living room and they began their noncompliance routine. Nancy, the legislative aide, was right on top of it. Many times the chief legislative aides are actually sharper than the politician they work for. I don't know if this was the case in this instance, but she was pretty sharp. Over her half-moon glasses, she queried these two officials: "Gentlemen, is it possible that these regulations were written by bureaucrats in Richmond who could not have conceived of an operation like the Salatins are running?"

I wanted to jump off the sofa and hug her. What could they say? She nailed them. They squirmed. I then handed them a wonderful document created by a couple of upper level biology majors at our local James Madison University for a lab project. One of the girls' families had been longtime customers of ours. Their project consisted of comparing total bacteria counts on our chicken and the numbers on supermarket birds. As part of the project, they wanted to get swab samples from the live birds, then right after processing, then from the retail location, to try to see where contamination really occurred. Was it production, processing, waiting in inventory? Good research.

No industry official or farmers would allow them to set foot on a production farm, nor would anyone let them into a processing facility. They finally settled for the only samples they could get: the retail level. They came and swabbed a couple of our chickens and bought a couple from the local supermarket and swabbed them. They cultured the swabs and results were:

Supermarket: 3,600 colony-forming units per milileter to the second permutation

Polyface: 133 colony-forming units per milileter to the second permutation

I honestly don't know all the scientific notations on the report, but I did know enough to realize that our chickens were infinitely cleaner—25 times, to be more precise. I shoved the report over to them and they looked at it.

Now wouldn't you think that if these guys were really interested in food safety they would have turned cartwheels over finding a model that really produced food that much cleaner? I mean, you'd think they'd get on the horn and call their bosses: "Hold the presses. Hold the presses. We've got the answer to everything right here. A new model promises to eliminate all food-borne pathogenic problems. Let's rewrite the books, folks. It really is possible without irradiation and chlorine. Let's print up a medal for this dumb farmer to recognize his culture-saving accomplishment."

You and I wish. Yeah, right. No, they glanced at it, gave it back, and immediately started shadow boxing again: "Well, pathogens aren't the only thing we're concerned about. We're concerned about the water quality, the temperature in the chill tanks, ambient temperature in the bagging room..."

I finally couldn't stand it. "Who cares about any of that if they're clean? If dunking them in the toilet made them that much cleaner, who cares? Are we after competency here, or infrastructure?"

Finally, after much hemming and hawing, they agreed to let us submit a written justification for our system, to argue its compliance with the federal guys. The reason the federal guys were involved is because the inspectors' salaries are paid half by the state and half by the federal government. And a state must have "equal to" status, meaning it cannot be less tolerant than the federal allows, in order to get cost-share inspection funds. It's more complicated than that, but to wade through all the nuances would take another book, and neither you nor I need to get bogged down in that minutiae. Just enjoy the story.

We submitted our argument for the no-walls deal and pointed out that "sanitary and unadulterated" are subjective terms. As long as we complied with that, we were in compliance and the law did not in fact give the regulators the power to further define what kind of processing facility, furniture, or appurtenances were necessary to create a sanitary product. That if it was, in fact, clean, then it was, in fact, clean. We took a strong position that if we complied with the government's regulations, our chickens would no longer be as pristine as they were. Compliance would in fact adulterate our product.

Amazingly, the feds agreed with us, and we won that battle. But dear folks, I want you to understand what emotional and physical energy all this cost us. The late-night conversations Teresa and I had when we could have been reading to the children or getting some sleep. The constant mental pressure to wonder what was going to happen.

Here's the point. We had ecstatic customers who would drive 200 miles to get this chicken. The whole model generated enough revenue to let us be truly viable, fulltime small farmers. Everyone was happy...except the chicken police. I say if you want to come to my farm, look around, smell around, and voluntarily opt out of government-sanctioned food, then our transaction is not a government incident.

As a postscript to this story, be assured that after our showdown, these same officials visited other farmers just like us in the state and tried to shut them down. They successfully shut down some farmers who didn't have enough savvy to stand up to them. Other farmers held their ground and the bureaucrats backed down. But the point is that even after we proved to the government agents that their language was wrong, they continued to spout these same lies to others. These people don't just stop when they lose.

They just take another paycheck from the taxpayer's pool and continue about their business as if the loss never happened. And they certainly didn't tell anyone in the system about a breakthrough model that could offer 2,500 percent cleaner poultry to the American consumer. They snuffed that light out before the match could get to the candle. At least we were free to continue serving our loyal customers.

Just to illustrate how bureaucrats run with rules, I will quote from a Feb. 3, 2005 letter from Pam McFarland, compliance supervisor, Illinois Bureau of Meat and Poultry Inspection, outlining the requirements for the poultry exemption:

"Poultry raisers with respect to poultry raised on their farms or premises (a) if such raisers slaughter, eviscerate or further process not more than 5000 poultry during the calendar year for which this exemption is being granted; (b) such poultry raisers do not engage in buying or selling poultry products other than those produced from poultry raised on their own farms or premises; (c) such poultry or poultry products are slaughtered, otherwise prepared, sold or delivered to the consumer on or from the premises for which the exemption is given; (d) such slaughter or preparation shall be performed in sanitary facilities, in a sanitary manner, and subject to periodic inspection by Department personnel; (e) persons desiring such exemptions shall submit in writing a request to the Department. The exemption shall be effective upon written notice from the Department and shall remain in effect for a period of 2 years, unless revoked. Adequate records must be maintained to assure that no more than the number of exempted poultry are slaughtered or processed in one calendar year. Such records shall be kept for one year following the termination of each exemption. Any advertisement regarding the exempt poultry or poultry products shall reflect the fact of exemption so as not to mislead the consumer to presume official inspection has been made under 'The Meat and Poultry Inspection Act' [650/5(b)]. "

This is obviously far more than what is in the federal exemption. At what point do bureaucratic protocols become law? While all this sounds fairly innocuous, it is not. Consider now the official Illinois "Review of Exempted Poultry and Rabbit Raisers" sheet. Imagine the power of the bureaucrat who can come onto a small farm and fill out the answers to the following questions:

1. "Is the procedural document acceptable? Yes No"

Did you see anything about a procedural document in the legislation above? No, it wasn't there. What happens is that the legislature creates enabling legislation, which is quoted above. Then the bureaucrats create templates and procedures. Suddenly the fairly innocuous legislation becomes a license to create mountains of paperwork. Be assured that the inspection agency will not help the farmer write a procedural document. The government agents will not provide a template.

The farmer, who's trying to juggle keeping the chickens alive, finding local grain, keeping rats out of it, designing a catchy logo, and answering the phone from interested neighbors who want his chickens, must stay up nights trying to create a document that the politicians never conceived. Many bureaucrats enjoy sadistic pleasure watching farmers try to comply with these kinds of things. Look, I know what is clean, what is cool, what is hot, what stinks, and what is ugly. Usually if clean, cool, hot, smelly, or ugly is handled correctly everything will be fine. And procedure doesn't mandate integrity anyway. Now on to the next line.

2. "Valid water potability certificate? Yes No"

Here again, the requirement has no basis in fact. When our farm test revealed chicken 25 times cleaner than supermarket chicken, we had some choliforms in our well water. The government has a zero tolerance for any choliforms. In fact, the test doesn't even indicate numeric level. It's just a pass/fail test. If these were deadly, nobody in history would have survived drinking out of streams containing deer dung and mouse urine.

I have a neighbor whose sons went on a missions trip to South America and were the only ones of a 25-member U.S. delegation to completely escape sickness. At their farm, they drink unfiltered spring water. My wife's grandmother lived to a healthy 100 years and 6 months old, drinking spring water all her life. We called it "crab water" because you could see little things crawling around in it. Immune systems are just like muscles. If we never exercise our immune system, it becomes lethargic. I drink out of mountain streams all the time.

Last year I had a siphon go down on a pond and in trying to restart it, sucking, the gunk in the pipe gushed into my mouth and I swallowed a couple mouthfuls of leaves, snails and who knows what else. Within 10 minutes I threw up and that was that. My body rejected it because it was an overdose. But little doses are like vaccinations. We've had people with all sorts of environmental allergies eat our chickens processed in unacceptable water, and we've never had anyone have negative reactions. These same people get rashes and terrible headaches because of the residual chlorine left on supermarket chicken. But even with all this said, I'm willing to have a water test. We did, and then installed an ultraviolet light and became compliant. Even I pick my battles. Next question on the form:

3. "Sales records reviewed? Yes   No"

What's the point? Anyone willing to lie about their numbers will fabricate sales records. Anyone who has dealt with these bureaucrats knows that they don't have a clue what they are looking for. Anyone can just make up records, show it to these agents, and create a completely fabricated record.

The whole process assumes that the farmer can't be trusted, and then creates a paper trail that only functions if the farmer can be trusted. What hopeless schizophrenia. And beyond that, isn't this a little invasive? Would you want to open up all your sales records to some bureaucrat? Now the next one:

4. "Was the operation reviewed during actual production? Yes   No"

I don't have a problem with this one except that the law does not require it. And what if the bureaucrat wants to see it in operation when it doesn't suit me? Perhaps I'm processing two days of the year—many small operators do exactly that. If a bureaucrat wants to be ornery—and many of them do—he'll just say it doesn't suit to come out that day. Perhaps the farmer is working a job in town and can only process on Saturdays. Bureaucrats sure don't want to work on Saturday.

The bottom line is that even something as simple as having a chicken policeman present during processing can become a huge problem. These requirements, of course, are formulated in an inspection context that assumes massive slaughter houses operating around the clock every single day. In that climate, of course, inspectors can pop in and out at their convenience and always see something going on. But on our small farm, no processing goes on most days of the year. Now the next question:

5. *"Is bleeding/picking/evisceration done in separate areas? Yes No"*

Again this assumes an industrial setting. Think back to the old hatchet, stump, scald tub, and board table set on some saw horses. The whole process occurred within a few feet, right in the same area, with children frolicking after the chickens doing their headless gymnastics. What is an "area?" Is that square footage in the back yard, or is it square footage in a building, or in a shed, or in a garage? These are the kinds of things that all chicken police interpret differently.

What inspectors love to see are impermeable partitions between the kill, pick, and evisceration areas. When a chicken dies, it goes through muscular contractions and sometimes shoots manure out its vent. Quite a dramatic finish, to say the least. In the picking process, of course, feathers can flip out and land on carcasses being eviscerated. But just like anything, on an extremely small scale, these possibilities that pose real sanitation risks in an industrial setting are just not problems.

We've had literally thousands of people come and process with us and they never complain about filth, ugliness, or lack of sanitation. If a feather happens to flip over onto a chicken I'm gutting, I just wipe it off. In an industrial facility, that feather might ride the carcass for some time before being removed. And what is "separate?" Does that mean a curtain, a concrete wall, fiberboard, or just a spatial separation? How a bureaucrat interprets that can mean the difference between no cost and

$20,000. And that is the insanity of these things: the raw subjectivity that puts the farmer at the mercy of a government agent's whim. It's no wonder people go postal.

6. "Facilities and Sanitation. Acceptable Unacceptable"

Here again, we have a totally subjective evaluation. And just what are acceptable facilities? Can you see the prejudicial slant on this form? And the incredible latitude it gives the field agent, indeed the department, to demand any sort of infrastructure they deem necessary? Remember, these agents live in a world of short cuts and intrigue. A world where large businesses try to shave hundredths of a penny off per chicken because the product margins are in fractions of cents.

And they live in a world where thousands and thousands of animals are killed each day. Literally tractor trailer loads of blood, feathers, and intestines must be carted away. Workers operate in surreal rooms filled with fecal water, cavernous concrete bunkers with bright overhead lights illuminating thousands upon thousands of dangling carcasses. Every day. Day after day after day after day. For these agents to step from that world into one that involves the children, pet cat, pet dog, robin chirping noisily from the shade tree, Dad, and Mom, and five neighbors stopping by to pick up their chickens... well... it's just about impossible to understand.

7. "Running water provided? Yes No"

Now what is running water? My grandmother always said she had running water: "We ran to the creek with a bucket, dipped it in, and ran home with it full," she'd laugh. Somehow, I don't think that's what the modern American chicken police call running water.

But in a serious vein, what if it's gravity fed from a bucket or tank mounted in the ceiling of the processing shed? What is sacred about running water? Now granted, we have piped water in our facility, but that's not how we started. The first year we processed chickens, we set up a wood fire under a pot for scald water and set our table up in the yard. I dipped water from a bucket to splash over the chickens and we went through 50 in about 4 hours. Today we do 200 in an hour. But our birds today are absolutely no cleaner than those first ones we did standing on the grass in the backyard.

If water is clean, who cares if it moves or not? This requirement alone could make a farmer trying to birth an embryonic local pastured poultry business suddenly have to spend thousands of dollars in order to sell one chicken.

8. "Hot water provided? Yes No"

For the uninitiated, all industrial inspected facilities must have foot-operated hand washing stations. Workers need to be able to wash their hands without touching a spigot or knob. And with thousands of square feet of surfaces to clean, lots of hot water is necessary.

But if I'm processing one chicken in the back yard for my mother-in-law across the fence, why do I need hot water? Maybe I'll gut the chicken in the kitchen sink. And for anyone grossed out about that, let me assure you that I would much rather put my hands in some fresh chicken guts than handle supermarket bologna or salami. I can scarcely look at that stuff without retching.

When we clean our little facility, we just take a 5-gallon bucket down to the hot water heater, draw off a bucket full, and carry it out to the shed. We wash down the two sinks and the chill tanks with that bucket of hot soapy water and it's plenty. Does that fill this "hot water provided" requirement? Probably not. Would the chicken police demand an on-site water heater? If so, isn't that kind of overkill for one chicken? Do you begin to see how absurd all this is? Now onto the next question:

9. "Dressing/Evisceration. Acceptable Unacceptable."

That's clear as mud, don't you think? Dear people, do you know how many ways there are to eviscerate a chicken? Whole books have been written about technique. This is as diversified as the numbers of mentors who ever dressed a chicken. As a

43

professional chicken gutter, I'll just throw out some of the questions I could ask about this:

- Do you pull off the head or cut it off?
- Do you electrocute or just slit the jugular?
- How long before scalding?
- Do you loosen the crop, cut it off, or leave it?
- Do you cut the vent out first or last?
- Do you cut out the oil gland or cut the tail clear off?
- If the crop breaks off, which way do you pull it—front or back?
- Do you break the large intestine before cutting out the vent?
- What if you break the gall bladder? What do you do?
- Do you separate the heart and liver?
- What do you do with the gizzard?

Obviously, this procedure is remarkably subjective. Some chicken police will not allow table top evisceration—they require shackles. This suddenly sets up a whole different set of logistical questions and techniques. With no more to go on than this section's title, the farmer again is completely at the whim of the bureaucrat. This is not 2 + 2 = 4. It's gray blob plus gray puddle equals nondescript subjectivity. Which brings us to the next section:

> *10. "Equipment and containers. Acceptable Unacceptable."*

Talk about wide open. Is a wooden bowl unacceptable? How about Tupperware, aluminum stock pot, earthenware, galvanized tub? Before we found our stainless steel chill vats, we actually used cattle troughs as chill tanks. We'd clean them down with hot soapy water and fill them with clean well water. They worked great, but I'll bet a sniveling bureaucrat wouldn't accept them even if the chickens were 10 times cleaner than birds at Cargill.

Furthermore, equipment for what and containers for what? This can easily include containers for feathers, blood, and guts. At our local custom beef slaughter facility, the cow police would not allow the butcher to put the guts in our barrels. The barrels had to be especially made with lead-free epoxy coating and marked, in equally specific paint: "INEDIBLE."

Didn't matter that we were just bringing the guts home to compost, in our trailer, from our animals, in our barrels. What did they think we were going to do? Feed it to our kids for lunch? Give me a break. I stood there and put the guts in there, do they think I don't know what's in there? Yeah, I'll get home and forget it was guts and think it's T-bones. Hey, wife and kids, come and enjoy some grilled pooh cord. How about some barbecued stomach lining with cud-chewed ooze attached?

In our backyard chicken processing shed, we use 5-gallon buckets for the guts. But they aren't marked "INEDIBLE." The contents don't really look very appetizing to me. I'm sure these buckets wouldn't pass unless we wrote, in some sort of special high-dollar indelible ink, "INEDIBLE" on the side. What a crock.

11. "Product presented to the customer:
Hot Fresh chilled Frozen."

What difference does it make how my customer wants her chicken? We've sold chicken with the heads on. What business is it of the government's how my customer wants her chicken? We'll do it however the customer wants it, thank you.

12. "Packaging area. Acceptable Unacceptable."

Oh boy, here we go again. The prejudicial notion that we have a building with a huge business going on. What if my packaging area is a lawn chair with some plastic bags on it next to the bucket of cold water I drop the birds in? I pull them out, stick them in the plastic bag, and off they go.

Or in our case, many of the birds aren't packaged at all because our neighbors come on the day of processing and pick them up fresh right out of the chill vats. They throw them on a bed

45

of ice in a picnic cooler and take them home to cut up or bag themselves. We got in trouble with the chicken police because in such cases we were not slapping a label on the carcass.

Finally the bureaucrats allowed us to put a label on the cooler. What's the deal? Don't you think these customers know where their chickens came from? And if someone wants to allege they got dirty chicken from somewhere, why would you trust where they said they got it? Maybe the company president made his ex-wife mad so she alleges she got junk chicken from that company. Labeling doesn't do squat in these kinds of circumstances.

Why have a packaging area if you aren't packaging? And what is an acceptable packaging area, anyway? I guarantee you it isn't a lawn chair holding a box of plastic bags. It will be a room, properly screened, properly sanitized, that could easily cost thousands of dollars. To package one chicken. And this area of course is completely separate from the evisceration area. It's just nonsense. If you go back and reread the enabling legislation, back before we started this review report, you will not see any of this stuff in there. Let's press on.

13. "Material: Plastic bag   Paper Wrap   Box Customer Containers."

Just more invasion that's none of the government's business. How did the agency get here from a simple sanitary product? It's beyond me. As far as I'm concerned, if a customer wants to drop her chicken in a dog food bag, who cares? We have some customers who buy our chickens for their dogs, for dog food. What would the agent do about that? Just leave the guts in and the feathers on—makes a great extra treat for the dogs, you know. Kind of tickles the throat going down.

14. "Where/how is product marketed? (i.e. signs, hand bills, newspaper, radio, TV, etc.)"

Again, why is this any of the government's business? At least they don't have an acceptable, unacceptable delineation on this one. Apparently, this is just for information. But do you see how condescending, how arrogant, these agents are? But during the review process, I, as the farmer, must happily supply all this information. Meanwhile, the chickens need water and I've got two customers on the answering machine that want product. But I'm stuck out here being nice to an invasive pest dedicated to helping me whether I need it or not.

15. *"Inedible   product   handling.   Acceptable Unacceptable."*

This is blood, feathers, skins, viscera. I have no clue what acceptable handling means. I assume it means containers or something, but who knows? Oh, that's right. I find out when the friendly chicken police arrive. They'll set me straight. Thank goodness.

16. *"Method of disposal: Burial   Rendering   Co. Pickup   Sanitary Service Pickup   Other."*

Notice what is completely lacking here? They don't even offer the best choice: composting. We've been composting chicken offal on our farm for decades, and it's still illegal without a permit.

Let's get this straight. In Virginia, it's legal to compost beef guts—as long as they are transported in crash proof containers. Sheep guts you can dump along the road. Nobody wants them due to scrapie. So even the slaughterhouses make you come and pick them up and you can dispose of them however you want to.

But chicken guts, now that's a different story. They can only be composted in licensed facilities. The industry is paranoid about chicken diseases, and has colluded with the duplicitous politicians to decree that no chicken or parts thereof may be composted without a government permitted facility. In our run-ins with the chicken police, they have mercifully let us slip through the cracks on this one.

If you bury the guts, rodents come along and dig them up. And they don't disintegrate for a long time. Any service that would come and pick them up will charge a hefty pickup fee. For one chicken? Give me a break. An idea I like—which we haven't done—would be to put all the guts in a floating tub on the pond. When the flies come and lay their eggs in it, the maggots fall out of the bottom of the tub and feed the fish underneath. I call it a Grub Tub.

See, when you're small all these wonderful micro-processes can be utilized to create circles of waste, regeneration, and renewal. But in the industrial set up, the streams of everything are too huge to handle in ecologically friendly ways. And in ways that are neighborhood friendly. As a result, these facilities are a menace to sanitation and the oversight agents naturally bring that mentality to local food systems.

That's the end of the review process. Obviously if a bureaucrat elected to take all these portions to a heavy-handed requirement, the first chicken would never be sold to the neighbor across the street. Every day in this great country, potential local food producers receive visits from government agents demanding compliance. And cowed would-be farmer-entrepreneurs, or agri-preneurs, as Ron Macher, editor of <u>Small Farm Today</u> magazine calls them, never get started. It's just too hard to be legal.

Chapter 4

Custom Beef

"I'm Bill from the Virginia Department of Agriculture and Consumer Services, Meat and Poultry Inspection. We've impounded your beef hanging at the slaughter house because an informant told us you're selling uninspected meat. We'll have to conduct an investigation."

The tall man standing in the front door of our farmhouse held a large metallic badge up by his face during this introductory monologue. I had come in for lunch, and Teresa joined me at the front door.

The week before, we had taken our entire year's beeves over to the slaughter house and the carcasses were now hanging in the chill room waiting to be diced and sliced into T-bones, ribeyes, ground, and the regular assortment of roasts. I knew we had done everything legally, so I figured it must be a mistake. Understand, that in those days, those twelve carcasses were a third of our annual farm income.

"What's the problem?" I stammered.

"It has come to our attention that you are selling uninspected meat. We will have to conduct an investigation to determine if that is true."

To understand the gravity of the situation, let me put things in context. The animals are slaughtered and then hung in a chill room for a week or ten days. Then the butcher begins cutting them into steaks and roasts. Those packages go into a walk-in freezer on wire racks to speed up the freezing process. Meat cannot be hung indefinitely without going bad. This is a highly perishable product. It had already been hanging for nearly a week when this public servant showed up at our door.

"How long will this investigation take?" I asked.

"As long as necessary to determine if an infraction has occurred."

"This isn't the first time you've done this, I'm sure, so could you give me a ballpark time?"

"Maybe six months. Whatever it takes."

I gulped. "In six months, this meat will be rotten."

"Too bad," he replied, completely unsympathetic. "If it takes that long, we'll just send it to rendering for dog food." That's one of the things that fries me about these people. They can just waltz into your business and be cavalier about destroying your livelihood because they draw their steady paycheck, have the power of the police, and the authority of the attorney general behind them. No apologies, no feelings. Just like the Germans leading the Jews to the gas chambers. Just like the U.S. Calvalry hunting down Indians. Just like Protestants hunting down Anabaptists in Europe. "Just doing my job. Just following orders." I am sick and tired of people who "just follow orders."

How much evil throughout history could have been avoided had people exercised their moral acuity with convictional courage and said to the powers that be, "No, I will not. This is wrong, and I don't care if you fire me, shoot me, pass me over for promotion, or call my mother, I will not participate in this unsavory activity." Wouldn't world history be rewritten if just a

51

few people had actually acted like individual free agents rather than mindless lemmings? Anyway, I digress.

"And what if we go ahead and cut it and package it anyway?" I asked, testily.

"We'll put 24-hour federal marshals at the slaughterhouse to make sure nobody touches it," he said. Each carcass was identified with a vibrant pink slip. It was equivalent to a police "Do Not Cross" tape to protect a crime scene.

"So how can we facilitate this investigation, to move it forward rapidly?" I asked, now fully aware and appreciative of the gravity of the situation.

"Give me the names and addresses of all your customers," he said, noticing the slightest hint of resignation come over my face.

"All my customers and their addresses!" I practically screamed. Digging deep within my soul for the conviction to fight city hall, I raised up on my hind legs and in no uncertain terms said, "No way. Not on your life."

And I turned around and went over to the telephone and called my senator. The senator called the inspection bosses in Richmond. Then I called my delegate. He made the same calls.

I won't bore you with details, but what followed was about a week of hell for us. This compliance officer practically parked at our front gate. He would come in and talk to us. We would call our politicians. A couple of hours later he would come back in with new news from "Richmond." "Richmond says this, and Richmond says that."

This continued from Tuesday noon until Friday late morning, at which time he walked back to the front door and said, "Richmond has informed me that we will assume this never happened."

To which I responded, "Look, man, easy for you to say. Trust me, it will not happen again. And to make sure, we will have a meeting here of your bosses to make sure it doesn't happen again "

Don't stop reading; the story isn't over. It's rich, real rich. Stay with me on this one, because it's the best story I have. And the sweetest victory. To set the stage, the United States has three kinds of meat slaughter and processing. For the record, meat does not include poultry. They were separated sometime around the beginning of language, I think.

The three kinds of meat processing are:

1. Custom. This is when a customer brings in an animal owned by that customer to be butchered and taken home by that customer. The operative language in the law is "an animal of their own raising." In other words, the beauty of custom facilities is that they form the backbone of neighborhood abattoirs. They must comply with rudimentary sanitation regulations, but have very little paperwork, no testing, and enjoy low overheads. Each package they process is stamped "NOT FOR SALE."

Envisioned as a way for people who wanted meat processed for themselves to get it done in the community, the custom designation went through a winnowing process during the late 1960s and early 1980s until the back-to-the-land movement created more local business. During the 1960s, at the height of the industrial economy, cattle farmers did not even eat their own beef. Remember, this was the height of Infamil and Similac, when breast feeding was considered Neanderthal activity. Then along came the beaded, bearded, braless revolution, Woodstock and Hippies, and La Leche League was born. This movement carried with it a newfound desire to raise my own cow and pig and have it processed in a community abattoir; indeed, this movement fueled a renaissance of sorts in the custom processing business.

And farmers like us, who began selling to these newly aware societal opt-outers, began hauling more and more animals down to these facilities because often they were the only processing game in town. Since our customers did not actually raise the animal, everyone in the system agreed that as long as the animal entered the kill chute with a name on it, the custom designation was okay.

In other words, it turned out that "of their own raising" was just too hard to police. What did that phrase mean? How long must I own it until it is of my own raising? Since the law did not specify how long an animal had to be owned to qualify, the default time period was zero. If I showed up to the custom butcher and told him that Jim Smith owned steer ear tag number 53, then who was to say Jim Smith didn't raise that animal? It was just too hard to police.

This wonderful loophole, still used today, is the backbone of what is known as the freezer beef trade. This denotes the volume sale, where someone will buy a quarter, half, or whole animal. If a customer buys a quarter of an animal, the farmer pairs that customer with others so that when ear tag 55 goes into the kill chute, paperwork clearly denotes that it is owned by four people, each getting a quarter. Those people then tell the butcher how they want their portion cut and packaged (Custom Processed), the packages are stamped "NOT FOR SALE" and the customer picks up the packages, takes them home, puts them in her freezer and later eats them for dinner.

The safety valve in this transaction is that if I bring an animal in, live, that I've raised, it's my responsibility whether that animal is sick or not. If I bring in a diseased animal, the butcher processes my diseased animal, and I get sick from my diseased animal, then that's my responsibility. Finally, a little sanity in the system. But wait, we won't be sane for long.

The owner can't sell those packages to anyone. She can give them away. But she absolutely can't sell them. That is a big no-no. It's what's called selling uninspected meat. Which brings us to the second butchering option.

2.   State inspection.   About half the states have a state-run inspection program that is similar to the federal but allows only intrastate sales. Nothing across the state lines. An inspector is on the premises at all times to sniff and visually inspect every animal. Infrastructure requirements are almost identical to federal and paperwork is less. But the packages are not marked "NOT FOR SALE" which means the meat can be sold to restaurants, retail establishments, and cuts can be sold after slaughter. All within the state.

3.   Federal inspection.   This is considered the highest form because it requires the greatest amount of infrastructure and the greatest amount of paperwork. As you might imagine, the average size of the facility increases as we move up the nomenclature. In common language, only state and federal are considered to be "under inspection." When an official says meat has been butchered "under inspection," he means either state or federal. Custom does not require an inspector on premises and that is why animals done there are not "under inspection." Even though custom facilities are inspected a couple of times a year, and have definite infrastructure requirements, only when an official inspector is looking at the animals is anything considered "under inspection."

Of course, meat processed under federal inspection can be sold interstate and internationally—assuming import-export requirements are met. Inspectors hold a lot of clout. They can shut down a facility at any time. The owner of the facility we use often meets me at the door with this greeting: "I don't own this plant." What he means is that he has all the risk, has all his life's equity and sweat tied up in that plant, but in the final analysis, he's not the one who sets policy or decides whether it's open or not. It's that federal official who draws a steady paycheck regardless of what happens.

When the bureaucrats pink slipped and impounded our beef at the custom butcher, they thought we were getting it processed, bringing the "NOT FOR SALE" packages home, and then selling them by the piece to our customers. In other words, selling them after slaughter. That is totally illegal. Once they went back over and looked closer, they realized that each of our animals had a name or names on it. Now they were in a pickle. They couldn't just admit they had done this in error, so in order to save face, they changed their interpretation and asked me to write an affidavit that I wouldn't sell the meat after slaughter.

I was happy to do that, because I hadn't done it anyway. It was like agreeing to do what I'd been doing anyway. I had no problem putting that in writing. That allowed them to pull the pink slips. But the reason that wasn't the end of the story was that they issued me a new interpretation: No matter what agreement we had with the customer, and no matter if the name was on the animal before it was killed, if the money exchanged hands after the slaughter, it was automatically an after-slaughter sale and therefore illegal.

That's why I had to have a showdown with them. I went to our custom butcher and talked it over with him. His freezer was full of beef from farmers who were selling it exactly the way we had done. But the government hadn't come after them. He had never heard of such foolishness.

In fact, several times I asked who their "informant" was. After all, isn't the accused supposed to be able to face his accuser? Oh, that was confidential information. We never found out for certain who it was, but through a series of actions, we deduced that it was another farmer who had an ax to grind and concocted this allegation to cause us problems.

Unfortunately, too often the person who rats on the black market cheese maker is the one who is struggling to comply with the regulations. Or the rat on a bootleg raw milk dairy is a neighboring dairy farmer angry that someone has figured out a way to triple the price of his milk. That's the way these things go.

The important thing to remember is that even though many, many farmers in our county were selling through this loophole that had been used since the earth hardened, we were singled out. And then the government agents concocted this interpretation that a sale could only happen when the money exchanges hands. Now folks, you know that isn't true. Binding contracts are written all the time. If you have a contract on the house, it's sold. And if you don't honor that contract, you can end up before a judge.

Each of our customers had sent in an order blank. We had a name on each animal. They agreed to a price. They agreed to a date. If someone drove up to the farm and wanted an animal already designated on our order blanks, I would tell him that it was already sold. Didn't matter that the money hadn't changed hands yet. When the deal is signed, sealed, and delivered, the money exchange is almost academic.

The pink slips came off, but our problem remained. How were we going to sell these animals in the future with this new interpretation? Our conundrum was simple. Freezer beef prices are usually based on carcass weight. The reason is that the dress-out percentage varies substantially from animal to animal. Let's say we have a 1,000 pound steer that dresses out 50 percent—that's a 500 pound carcass. If another one dresses out 60 percent, that's a 600 pound carcass. Customers don't eat the hide and guts; they only eat the carcass. But one yielded 100 pounds more than the other. That's a big difference.

If we sell by liveweight, these yield percentages don't enter into the equation. If we price it to protect ourselves from the low yielding animal, then the customer who gets the high yielder will pay an unfair premium. If we price it to assume a high yield carcass, we'll get financially hurt with the lower yielding animal. The only way to make it fair is to price it by carcass weight, but we don't know the carcass weight until the animal is dead. Hence, we didn't have a way to create a fair price based on a liveweight.

Finally, our state delegate, who happened to be the minority leader in the General Assembly at the time, asked if a waiver would be in order. He asked if our customers would be willing to sign a waiver absolving the government of liability and recognizing that they took full responsibility for what they were buying. I assured him they would be glad to do that. Our folks trusted us completely, and understood very well that these battles had nothing to do with food safety.

He drafted a waiver and submitted it to the state attorney general. The attorney general said waivers are a joke. They don't stand up in court and they aren't worth the paper they're printed on. End of discussion.

We set a date for a meeting with the top state inspector and the federal officer responsible to oversee the state to make sure the state did not violate the "equal to or more stringent" requirements of the federal codes. The federal bureaucrats use "equal to" language to hold states hostage to federal mandates. If a state were able to grant more freedom to its citizens than the beltway crowd thought prudent, people in other states might like the new opportunities and vote themselves unimagined liberties. The feds can't abide such a notion. It would ruin our socialistic country. People owning their own bodies? Their own decisions? No, that must never be.

When the bureaucrats showed up, they were half an hour early. I went out and told them they were half an hour early and could not get out of the car. My people were not here yet. My attorney and my delegate's aide came, along with our homeschooled children and apprentices.

When our people showed up, I waved out to the bureaucrats that they could come into the house. Our children were right behind me as I welcomed them into the house with a brief introduction that this was like a field trip to see how the government works. I assured our children that these officials were here to help the citizen and to make sure farmers could continue to succeed. To insure that Americans would enjoy freedom of choice in their food, and that righteousness would prevail. We had a special gallery set up in the living room for our children and apprentices.

Of course, this introduction put the bureaucrats a little on the defensive, which was exactly what I wanted. I tell folks all the time that you don't have to shoot these agents, but you don't just have to roll over either. Have the self-awareness and confidence in your moral high ground to maintain your dignity in their face.

The opening salvo came from the aide and our attorney, who wanted to know the history of this new interpretation that no matter what transpired before slaughter, if the price was based on a dead weight, the sale was automatically an after-death sale and therefore illegal in a custom facility. The officials assured them that nothing prompted it at all, it was just Monday morning and time for a new interpretation. They said it was unfair for us to slip through a loophole that allowed a $30,000 facility to process beef when Kroger and Wal-Mart had to use $1 million facilities.

"These retail outlets are on us every day wanting to know why we let people like you get access to the market when they have such a high cost in compliance."

"Well, we're not in the same game," I said. "Our customers received an order blank in the mail six months before these animals were ready. They ordered and signed their names that they would take a whole, half, or quarter. They had to think ahead half a year to do this. And they need a big freezer to put it all in. And they are going to be cooking this stuff in their kitchen. They have to come out to the farm to pick it up, so they see these animals and they know us. These folks know what we feed, what we don't feed, and how these animals are raised.

"The folks you're talking about that are putting the squeeze on you don't sell beef this way. Their customers can come in at 5 p.m. and buy a steak on a whim. They don't have a clue where the meat came from, who grew it, or what the cow ate. They have no relationship whatsoever with this meat. And to be sure, we have absolutely no problem with them playing in that game. But for these retailers to even think that we are in the same game is ludicrous."

The officials responded that their retail constituency did not see it that way. Meat was meat, and they believed we had an unfair advantage in the marketplace if we were going to continue to use low-cost neighborhood abattoirs. They didn't have an answer as to why they had singled us out but hadn't pursued any of the other carcasses hanging in the same chill room being sold exactly the same as ours.

Things degenerated from there for awhile. Finally, in a rare stroke of genius, I asked what proved to be the pivotal question of the day: "If you were me, what would you do?" And I shut up. Everyone shut up. Deathly silence.

I learned a lot from that experience. Many times, in a rather heated disagreement, we can bleed off the heat by asking the other side, sincerely, what they would do. Suddenly they can't just complain. In formal debate (I was on the debate team in both high school and college, and still consider that extracurricular activity to be the most valuable educational investment of my life) the affirmative side must win two elements: the case and the plan. The case is the complaint, or the problem, with the status quo. The affirmative is always asking for a change from what is.

But in order to win, the affirmative must also have a workable solution, called a plan. If the judge agrees that the status quo stinks, that the complaint is valid, he can't vote for the affirmative until he also buys the plan. Often the affirmative will win the case, but lose because the negative team shows that the plan either won't solve the complaint or has such serious disadvantages in its own right that the cure is worse than the disease. That is a good rule for life. Anyone can complain. That's easy.

But what to do about it—now that's the big question. And so I just put it to them. I was tired of hearing them complain about us. I wanted a solution. And then we all shut up. The silence was long. I will never forget it.

Then the federal guy, newly energized by being asked to help, offered this suggestion: "Why don't you charge something by the head—an amount that every animal will generate. Then adjust the price based on carcass weight by calling it shipping and handling. That way you satisfy us with a before slaughter sale but you also can customize it to carcass yield by adjusting on a carcass per pound price based on shipping and handling."

We all sat there, stupefied for a moment. So simple. It was profound. How could we have possibly been so uncreative at insuring that our beef would be safe for our dear customers? Forsooth, our customers could now eat in safety. Glory, glory, and my, my. The man was a genius. The meeting was over. We agreed that we would do that; they agreed to leave us alone, and we parted amicably.

The next year, we had a new pricing structure: our beef was $1 per head; half 50 cents; a quarter was 25 cents. But that shipping and handling was rough--$2 a pound carcass weight. We explained the whole charade to our customers and they thought it was a hoot. Of course, they played right along. Most folks get a deep soul level satisfaction at pulling a fast one over the authorities.

You know the funny part? Virginia collects sales tax on food and merchandise, but not on services. We were no longer in the beef business; we were in the service business. Instead of collecting and sending thousands of dollars in sales tax to Richmond, we sent a few pennies. After all, you have to be consistent. They made the rules; we were just playing their game. Isn't that hilarious? Here the best officials the state and feds could throw at us had succeeded in bilking the capital out of thousands of dollars, and saved our customers that money as well.

It was truly a great day for all. After several years we decided we should be more afraid of the sales tax vampires than the food police. We reverted to our pre-showdown days of just selling by the carcass weight, and no one has said anything. All this hassle, attorney's fees, mental anguish, and brouhaha, for what? Nothing really. Except a great story to tell. And hopefully a lesson learned. The lesson being that bureaucrats are not reasonable; that they have no heart; and that they are more stupid than they can possibly imagine.

I'd like to conclude this story with three points. The first is that if the custom product is safe, why can't we sell it? If it's safe, it's safe. And if it's safe to eat when your name is on the animal before slaughter, then why isn't selling a package from that animal also safe? I can give away those packages to people who have never met me and don't know me from Adam. I can use those packages at the local school for the kindergarten picnic. I can cook those packages for a neighborhood picnic.

But if one of those neighbors, one of those kindergartener's parents likes it and wants to buy a pound of hamburger from me, I'm headed to jail because the package is stamped "NOT FOR SALE." What is it about exchanging money for the package that suddenly makes it a hazardous substance? Unsafe food? When looked at this way, the thin veil called protecting people as a justification for all these regulations just melts away. I can assure anyone that those packages from our little custom facility that only slaughters a couple of days a week and operates on family labor are every bit as clean and safe as those that cost half again as much to be processed from the federal inspected facility.

You could eat off the floor of our custom facility. The inspectors made new rules demanding that every kill floor had to have a squeegee to push the blood down into the drain hole. That might be necessary in facilities processing thousands of animals a day. But in ours, where they do a couple per hour, such infrastructure is a complete waste of money. No matter, they had to buy one. It sits in the corner and has never been used. But they are under compliance. Oh, thank goodness.

In our culture, hazardous products carry prohibitions on both buyer and seller. Without getting into a full-blown discussion about drugs, the point here is that they are not only illegal to sell, they are illegal to own or use. But in food, the prohibition is only on the seller, not the buyer. And not the user.

If I talk a person into selling their "NOT FOR SALE" package of T-bone steak, I'm totally legal to buy it. And I can feed it to my children, their friends, and my neighbors. The only liability is on the poor farmer that sold it to me. My question is if it's really a safe food issue, it should be illegal to acquire the hazardous product as well as to sell it. It's either hazardous or not. Don't give me this pabulum about food safety. This is not about food safety. It's about denying market access to appropriate-scaled transparent products, and thereby insuring de-facto market protection for the current big players.

What these regulations do is to insure that whatever money is made in tomorrow's food system, will be made by today's players. Upstarts are not allowed. It's the ultimate game of "I got mine and you can't have yours" and it's rotten to the core. The people being contaminated with food borne bacteria are not getting it from custom facilities; they are getting it from industrial-sized closed-door federal facilities where the government-corporate fraternity speaks the same language and colludes to make sure that whatever truly innovative answers exist to today's problems will never see the light of day.

Second point. Our attorney is a lifelong Democrat. The delegate was a Republican. Our state senator who went to bat for us was a Democrat. Our Congressman who entered the fray was a Republican. Without exception, we have found that these issues excite both sides of the aisle. To the Republicans, this is small business and entrepreneurship held back by meaningless regulations.

To the Democrats, this is about environmental farming and chemical-free food accessing the marketplace. As the alternative food movement continues to gain steam, I enjoy watching the liberals squirm when they find their freedom of food choice arbitrarily quashed by their partners in the government. Those folks that are supposed to insure fairness and equality for all the citizens.

And it's equally interesting to watch the Republicans squirm when they realize the collusion between the bureaucracy and Wall Street. The coziness between tax breaks and the seats of power. Corporate welfare. When tax-free bonds are handed out like candy to big players but little players get whacked on the nose if they have one "NOT FOR SALE" package on an invoice.

At some point, the government-loving liberals need to think about where their well-meaning government help is taking us, and the Chamber of Commerce conservatives need to appreciate the thrashing that environmental food entrepreneurs receive at the hand of Wall-Street and the USDA. All of us have plenty of learning to do.

Third and final point. At the time of this showdown, we had an apprentice here, newly graduated from the University of Colorado in Environmental Sciences. Totally committed to the government solution for everything, he was indignant at the way these bureaucrats were handling us. He knew our meat was great and that all this was an orchestrated attempt to destroy an up-and-coming player in the clean food movement. He was ready to fight.

In fact, on an extremely personal note, I will share that during one of the darkest mornings, when the inspector was camped outside our front door, I was praying and meditating in the field during chores, in great anguish of soul and spirit. My question: "What do you want, God? Have you blessed us with these customers and these environmentally-enhancive production models to just shut it down?" Realize that even though I have separated our PL 90-492 battle from our custom beef battle, they occurred at about the same time. That was a rough year. I have separated them in this book for clarity and because the issues are quite different.

And in the quietness of that moment, while I was moving broiler shelters, with the gentle chirping of the birds as a background, the answer came back: "For such a time as this." My mind immediately went to the Old Testament book of Esther, when the Persian king had been duped by an evil courtier to command the extermination of all Jews. Esther, a Jewess, disregarded royal protocol, disregarded her life, and was admonished by her loving uncle and guardian, Mordecai: "For such a time as this." (Esther 4:14) She conceived a plan, at peril to her life, to save her people, and it worked. In the many years since that day, I have taken great solace in that answer.

In fact, that gave me the courage, when the officials tried desperately to shut down our outdoor chicken processing shed, to look them in the eye and tell them, "When my customer carries his 90-pound wife up to my front door and says, 'The doctor says my wife will die if she doesn't get organic meat. Can you help me?' I am going to make sure she gets it. Now you folks can either make life difficult or easy. It's in your hand. You know as well as I do there is no food safety issue here. If you want to see what my pen can do from jail, then go ahead and take me there. But I am telling you today, that like me or not, like this facility or not, our customers will get food that is not adulterated by your stamps of approval."

Immediately, I saw the officials soften. I was not screaming. I did not raise my voice. I think they saw passion like they had never seen. I was trembling inside. But this is not just a business, it is a sacred calling, a sacred ministry, serving people who seek truth and are willing to travel dirt roads to get it. Americans have been acculturated to bow before officials, and to assume that if you have alphabet soup behind your name, you are a decent, reasonable, honest individual. 'Tain't so, dear friend. To the liberals, I say that for every greedy, evil-hearted business owner you will find an equal in the government; and to the conservatives, I say that for every greedy, evil-hearted tree hugger you will find an equal in the government.

So our apprentice was incensed. "Call the newspaper," he said. "This is ridiculous. The media should know about these guys. It's not fair."

I looked at him and said, "Tell me. If you didn't know anything about this, and you opened the newspaper in the morning to this headline: 'Food Inspectors Accuse Local Farmer of Selling Uninspected Meat' what would you think?"

He responded quickly, "I'd think some scumbag farmer—after all, their cows are melting the polar caps and will soon flood New York City—is preying on unsuspecting consumers and selling them trashy food." He paused as the reality of his paradigm broke in upon his consciousness. I think it shocked him.

"You have to pick the field of battle," I said. "We can't beat that headline. You and everybody like you thinks farmers don't care about anybody, that all business people are out to pull a fast one on those poor ignorant consumers, and that whatever the government officials say is gospel."

He had to agree, and it was a real epiphany for him. It opened the floodgates for some real cross-partisan discussions. We'd all like to think that truth will rise to the top, but that's not always the case. Oh, it will ultimately, but ultimately can take a long time coming.

Here we are, a few gray hairs later, lots of lost workdays later, selling our freezer beef and pork just like we did before the Richmond Cavalry descended on us way more than a decade ago. And Richmond is getting its sales tax money. All is quiet—for now.

Chapter 5

Bacon

"I'd like some bacon. Why don't you offer that?" The customer's question seemed reasonable enough. After all, we sold sausage, ground pork, tenderloin, spareribs, backbone, pork chops, fresh pork roasts—why not offer bacon? To the average person, bacon is just a special kind of pork.

But not to the government. Bacon comes from the belly of the hog. It's the muscle that holds all the insides in. It's the same muscle that has become elongated and stretched on the front side of many people.

To make bacon, of course, the hog must be slaughtered and cut up into pieces. The hind leg yields ham, the front leg yields shoulder, the back yields the loin, and the ribs yield...yes, ribs. The belly yields a piece a little more than an inch thick and about 15 in. X 15 in. Curing this piece of pork with salt, pepper, and brown sugar yields bacon. Bacon is not a fresh piece of meat; it's a fresh piece of meat cured in some way. It can be cured with smoking.

I love bacon, and I certainly wanted to sell it to this customer. My father-in-law had a curing house and when Teresa and I were younger, the family would get together in November and have a hog-killin'. We'd dress out the hogs in the morning and let the carcasses cool down on tripod poles during lunch, which was a veritable feast.

After lunch we'd cut the hogs into their main pieces—ham, shoulder, back, belly, ribs—and begin cooking the fat for lard. The hams, shoulders, and bellies would be carried up to large wooden tables in the curing house and liberally covered with the curing recipe—usually salt, black pepper and brown sugar. We would rub this mixture into the meat.

The curing mix combines with the meat juices as it permeates. Cold nights and warm days aid the curing process by making the meat weep a little bit and then stiffen. This movement helps the tissue absorb the cure. The thicker the chunk, the longer it takes to fully cure. Hams often take four months but bacon, because it's only a little more than an inch thick, can cure in a month. After accepting the cure, the meat is ambient temperature stable indefinitely. This procedure enabled meat to be held for long periods of time before refrigeration or freezing were available.

Common knowledge and common procedure in the countryside, home curing gradually fell into disuse with the rise of industrial curing operations, the supermarket, and the rise of government regulations. It is an extremely simple process, does not require much infrastructure, and is easy to learn.

Why not just go ahead and bring back the big pieces of belly from the slaughterhouse and cure it here on the farm like we used to? We had the ability, the desire, the market, the infrastructure. Only one problem: it was illegal. Even though we had a federal inspected seal on the meat when we brought it home from the slaughterhouse, as soon as we did anything else to that meat besides sell it, we were in the processing business. And processing meat without a license is illegal. Okay, so let's get a license.

Well, it's not really a license like a driver's license. The average person is used to a license simply being a standard of competency. If you can drive the car and pass the test, you have proven competency and the license is yours. Just imagine if in addition to proving competency, you had to own a car with a certain horsepower. Suppose you had to own a garage, where you could park the car out of the weather. Suppose you could only get a license if you had a car big enough to carry five passengers. Unfair, you say?

Not at all, when it comes to on-farm bacon making. We needed an inspected curing facility. The meat needed to be transported in an approved container. This is one we've never been able to really pin down. We've been told that it's illegal to take inspected meat from a slaughterhouse in a non-refrigerated vehicle. But if we're only an hour from the facility and we crank down the air conditioner, it stays plenty cold.

No matter. Must be a refrigerated vehicle. Let's say we spring the thousands of dollars necessary for a refrigerated vehicle so we can legally transport the meat home. Now it goes into a permitted facility on the farm for curing. One little problem: processing facilities are illegal in agriculturally zoned areas, because this procedure is not considered farming; it's considered industrial or commercial use.

This curing facility needs bathrooms and an office for the bureaucrats to do their paperwork. It needs impermeable walls and washable floors, so many lumens of light per square foot, and handicapped access. It needs temperature control, with 24-hour monitoring disks. And we need to calibrate all the thermometers monthly—that only takes a few hours a month. And each calibration must be duly noted on a form with the non-transferrable thermometer delineation in order to insure efficacy.

After hearing all that, I realized we couldn't afford to do that. So I asked the inspector for alternatives to the curing facility. He said if we had a retail establishment, we wouldn't have to be subject to any of these requirements. Okay, so let's have a retail facility. We already had a nice on-farm store complete with wood burning stove circa. 1920 to provide heat and ambiance.

A retail store must have approved bathrooms. Now we're back into septics. It's not allowed in an agricultural zone. It must have an approved parking area and commercial entrance, including appropriate site distances up the public road.

"Look," I said. "All I want to do is salt down a pork belly to make bacon to sell to this nice neighbor customer. "

He patiently explained that bacon is a processed product. It doesn't matter if it's only one pound or a million, the infrastructure requirements are there to protect the customer from food poisoning.

"Well, what if I smoked it rather than the salt deal?"

"No difference. When you take that meat home, it must be received into a licensed facility. A licensed facility has certain requirements: concrete, rebar, lights, bathrooms, handicapped access, temperature controls."

The important thing to remember here is that competency does not require any of this infrastructure. Folks around here have been curing pork for centuries, safely, in crude on-farm cabins and outbuildings. Smokehouses are everywhere.

I finally realized that curing wouldn't happen on our farm for a long time, so I began calling businesses that cure. The big guys aren't interested—they are part of a vertically integrated chain in which the company owns the hogs, the slaughterhouse, and the further processing facility. After many phone calls, I located a guy a hundred miles away who agreed to do it only after I convinced him that I wouldn't try to take away his markets. He was buying tractor trailer loads of hams from the industry and curing them with his own recipe, and had a very dynamic, successful business.

He told me one day he looked out in the parking lot and saw six government cars. He told the bureaucrats that he couldn't possibly earn enough money, and the American taxpayers couldn't earn enough money, to pay for them to be there. He built this business over his lifetime and can make your hair stand on end with bureaucrat stories. Now he keeps a shotgun in his office and when the bureaucrats roll in, he steps out onto the porch and lets off a couple of shots in the air just to annoy them.

One catch—he'd never cured without nitrites and nitrates. These preservatives keep the fat from discoloring during and after the curing process. They reduce the skill necessary for a good cure. He said he would try to do it without these noxious preservatives. And the price? Hold onto your hats--$1 per pound to cure and $1 per pound to package. You see, he did it reluctantly. This was the only legal facility in the state where I could get this done.

Since packaging is part of the processing, he couldn't cure it and then let us bring it home to package it. In order to be legal, the product must be under inspection all the way to the final package. Of course, we had to get him the fresh pork. Fortunately, he came weekly to the processing plant we used to slaughter the hogs. We arranged with the slaughterhouse to freeze the product for curing and hold it until he came.

Our volumes were so small it took awhile to accumulate enough product to justify turning on one of his curing rooms to do it. When the first batch was finished, he called me with the bad news that it was no good.

"It's yellow," he said. "You can't sell that."

I looked at it, cooked some, and it tasted great. The yellowing was due to the absence of nitrites and nitrates. I assured him that it was okay and our customers wouldn't mind. He shook his head in disbelief. Sure enough, our customers loved it—and still do.

The pork shoulder cure wasn't as successful. After throwing away about $4,000 worth of meat, we finally gave up. He could never get the process right for shoulder without the crutch of the nitrates and nitrites. We gradually cultivated a market for Boston Butts and that solved the shoulder problem. He was more successful with the ham, and although we lost some product initially, he was able to make the cure work.

When we sell a pound of bacon, we have more than $1 in butchering the animal—that goes to the slaughterhouse. Then we have $2 to the curing facility for curing and packaging. That's $3 a pound right off the top. When we sell bacon for $6 a pound, we have only $3 left to pay for the pig, the feed, the labor, and the marketing. And that ain't much.

The problem with all these infrastructure and paperwork requirements is that they are non scalable. They discriminate against the small operator because the big outfits have enough volume to spread the high cost over additional product.

To illustrate this point, I have a neighbor who makes old-fashioned pickles. She went to the expense of installing a commercial kitchen in her large home basement and attached two-car garage. She uses five-gallon stainless steal stock pots for cooking on two six-burner high-BTU commercial stove surfaces. She makes salsa, pickles, jams, and jellies.

For efficiency, she makes a batch of product utilizing all the cooking surfaces at one time and then makes a batch of something else. Because she has twelve burners, one batch may use six or eight stock pots. But the Food and Drug Administration inspectors do not allow her to call that one batch. Each pot is one batch. She has to check Ph, cooking temperature, time start, and time ended on each pot and laboriously fill out the appropriate square in a chart for each of these. And she dare not miss one.

A large outfit cooking in a 1,000 gallon pot has the same paperwork as she does using a five gallon pot. The labor to hire a full-time paper-filler-outer for the small outfit is 20 percent of the labor force. For the big outfit, it's less than 5 percent. And if you happen to make a mistake, the bureaucrats will move heaven and hell to punish you.

Case in point. A friend in Minnesota operates a wonderful small slaughterhouse. He has to probe the meat and fill out a chart every hour to document the chill-down. They have a square to fill out at 1 hour post-kill, 2 hours post-kill, etc. for so many hours until chill-down compliance is achieved. One day, an employee forgot to do a mid-time check. The square was just blank. But the preceding one and the subsequent one were both right on target.

The inspector saw the blank square and demanded that $40,000 worth of hanging beef be thrown away due to the noncompliance. Obviously, this oversight had nothing to do with food safety. It was just a matter of human error. This is how intransigent and unreasonable these bureaucrats are.

I had an interesting attorney visit recently who represents the largest fast food chain in the world. He is part of that famous revolving door between Congressional staff and large corporate offices. He said the reality is that the large processor has problems with over-zealous bureaucrats too. But they have attorneys like him, on retainer, who simply fix it. His job is to contact the appropriate supervisors at Food Safety and Inspection Service and call off the dogs, so to speak.

He said bureaucrats are bureaucrats and don't really care whether they are dealing with big operators or small ones (at least that's his perspective). But it looks like they give the big guys a pass because the big guys can afford to hire go-betweens that use their ability to entertain, buy, bargain, or cajole to garner concessions as necessary. The small processors can't afford to hire a full-time attorney to do this work because they can't spread the cost across millions of pounds. As a result, when the small outfit feels the brunt of bureaucratic unreasonableness, the only recourse is to capitulate, repent in sackcloth and ashes, ask, "What must I do to be saved?" and try to placate the officials.

According to him, it's not really a discriminatory application of the law; it only appears to be so. If the little guys would become big guys and hire people like him, they could enjoy the benefits of retained insiders to run interference too. In other words, the solution for small businesses is to become big businesses and all will be well. He probably votes Republican.

In my years lobbying for scalable regulations in government, the common bureaucratic response is that we have to have a "level playing field." My response to that is that we aren't playing the same game. For them to invoke the level playing field requirement on a guy slaughtering 10 beeves a week in a facility attached to his house, when compared to a Tyson slaughtering 5,000 beeves a day in a 1,000-employee slaughterhouse is just ludicrous.

That's like saying it would be a level playing field to allow anyone to play football as long as it was done in an NFL stadium. No backyard pickup games, no high school gymnasium adjusted-rules stuff on rainy days. We're going to have a level playing field. If anyone wants to play football, it must be done in an NFL stadium. Now doesn't everyone feel empowered? Give me a break.

Why can't the rules be written to flex with the game being played? If we're in a backyard that's only 50 yards long, we say a touchdown happens in 50 yards, not 100. Who hasn't adjusted rules to fit circumstances?

The problem is that the rules become requirements of system rather than requirements of competency. If a touchdown is the goal, you can have just as valid a touchdown on a short field as a long one. If food safety is the goal, you can have just as safe a food produced in a home kitchen as in a Campbell's soup factory. If the goal is safe food, who cares what the infrastructure is as long as the food is safe?

These ludicrous requirements are an attempt to legislate integrity, and integrity cannot be legislated. A local delegate told me the biggest problem the poultry industry has in their big processing plants is that many of the workers don't wash their hands after going to the bathroom because that is not a part of their culture's hygiene. You can have all the stainless steel in the world, all the hot water and chill tanks you want, but in the final analysis, if you can't get workers to wash their hands, the product will not be safe. You simply cannot legislate integrity.

But somehow the government believes that if we just require enough squares in paperwork to be filled out, we can make a system foolproof. The big food poisonings and recalls are not happening in small outfits; they are happening in large outfits. No rules or protocols can insure honesty or care.

The result of all this is that for the small producer and processor, the nonscalable regulations create a discriminatory cost structure on the price of the product or service. When the small outfit enters the marketplace, its product or service is at a competitive disadvantage. People are always asking us why local meat and poultry is more expensive than what's at Wal-Mart that came from 1,000 miles away. It seems like it should be cheaper if it's sold across the fence to the neighbor and doesn't have all that transportation behind it.

The high cost of local food has nothing to do with actual costs. These costs almost always are a result of inappropriate regulations that preclude efficiencies. Like requiring a bathroom in the smokehouse.

The bottom line: most of what enters the marketplace is from large outfits. And these outfits constantly try to expand in order to spread the cost of regulations. The best way to eliminate competition in the marketplace is to stifle upstarts from ever seeing the light of day. And that is why so few pastured hog producers can offer bacon.

Just to prove my perspective, I will quote from an official document prepared by John A. Beers, Office of Dairy and Foods on May 9, 2003 as the official response of the Virginia Department of Agriculture and Consumer Services to a proposed amendment that placed goat, sheep, and water buffalo cheese on par with cow cheese regulations.

"The final regulation will ensure every person who sells cheese is competing on a level playing field. Such is not the case today. Currently, anyone making and selling cheese from cow's milk is required to meet the requirements of the final regulation. Persons making and selling cheese made from the milk of goats or sheep are regulated under less specific requirements contained in the Virginia Food Laws. Within the group of people making and selling cheese from goat's milk there is a division between those who are in compliance and those who are not. This situation leads to disparities between the three groups considering their respective cost of production. The current situation provides some individuals with cost advantages over others making the same or similar products."

Notice how he uses compliance and level playing field? This is typical justification for adding more burdensome and local-food-destroying requirements.

Perhaps the more telling portion of this same report deals with the official agency response to the charge brought up during public hearings that consumers should be allowed freedom of choice to determine what they choose to eat. Here is the official agency response:

"For individuals to make a choice implies that they have some basic knowledge on which to base a decision. The Department believes that the average consumer does not possess the basic knowledge to be able to determine if milk and dairy products are safe. Less than three percent of the population lives on a farm or has any understanding of the processes required to produce milk and dairy products safely. Consumers also lack basic understanding of risk factors involved with sanitation, production and processing methods, packaging, handling, labeling, and distribution. The average consumer does not question the safety of food products offered for sale but expects them to be safe. Consumers assume food products are safe because their experience tells them so, not because of their knowledge of food safety.

"Consumers also assume that products being offered for sale at farmers' markets or other places established by local government authorities are just as safe as products in grocery stores that come from inspected facilities. The fact that farmers' markets are sanctioned by local government gives people the impression that the food products offered for sale have been sanctioned by local government.

"Children are one group of consumers who have no choice. Children will consume what their parents or other adults provide them to eat. Children are unable to determine for themselves what is safe or unsafe to eat. Cheese and dairy products made from unpasteurized milk are associated with a high level of illnesses in disease outbreaks traced to dairy products. Children are often the victims of these diseases. "

Do you see the condescension here? The poor, stupid consumer is too pitifully ignorant to make her own decisions. And notice the ostentatious bleeding heart for children. Oh, they care so much for children. They have no problem with parents feeding their children Twinkies and soy milk for breakfast. Sitting for hours in front of the TV after school. Consuming half a dozen cans of soft drinks per week. Pumping kids full of Ritalin and feeding them feedlot beef.

If these guys really want to do something about children eating better, they would begin a crusade against half of what's in the supermarket. Directly quoting the department yields great insight into how these bureaucrats think. Clearly, they do not believe the American consumer is capable of making decisions. That should raise the ire of every freedom-loving citizen.

To take this one step farther, the report goes on to say that:

"The Department is unaware of any scientific evidence that supports the allegation that milk and dairy products sold directly from the farm are of superior quality or safer than commercially available products. Some individuals allege that because commercial dairy products are manufactured from the commingled milk from numerous dairy farms that they are more subject to contamination than similar products manufactured by a single farm. The Department believes that all milk and dairy products have the same potential risk of adulteration with pathogens or other organisms. The same steps needed to process milk and dairy products into safe and wholesome foods are necessary for both the individual dairy farm and the large commercial processor. Where food safety is concerned, smaller does not equate with safer."

Notice how the bureaucrats circle the wagons around the mega-processors? The prejudicial stance is too obvious to deny, and it's fascinating to see these viewpoints in black and white. Anyone who believes that bureaucrats love individual decision-making liberty and small, local food systems can see the opposite defended and encouraged in this report.

When coupled with all the reports that show centralization is the vulnerability point for pathogens and bioterrorism, such blind protection of the industrial food system could be described as evil intentioned, not just ignorant. Who says these folks don't have an agenda? Of course they do. The agenda is the systematic dissection of small, local food systems. Rather than just go to an empirical standard of so many parts per million of whatever, the government agents want physical structures and mountains of paperwork to level the playing field and prove compliance. It's obscene and un-American.

The tragedy is that these bureaucrats have equal sway over Republicans and Democrats. Both groups of politicians bow before these people as if their credentials and appointments came directly from God.

A couple of years ago I was in Richmond lobbying against a bill that would have given the Commissioner of Agriculture (the one who sat beside me and likened raw milk to moonshine) the authority to promulgate any regulation he desired on chicken farmers and chicken sales. These regulations, by statute, would have been exempted from any hearings or citizen oversight. A few of us got together and cried foul—no pun intended, and eventually succeeded in amending the bill.

Anyway, during the hearing process, people like me sat in the audience, out in the cheap seats. The industry insiders and the Farm Bureau Federation lobbyists, the lobbyists for the Agribusiness Council—all these folks were invited to pull up chairs and sit around the table with the legislators. The bald-faced, shameless cozying up that the elected politicos show to the powerful insiders is downright embarrassing. I come away from those meetings feeling like I've been dipped in some evil juice. I feel dirty and unclean, like I've become soiled with societal depravity.

If I pulled up a chair to that big oval table at the front, the Democrats, Republicans, and their industry buddies would call the state trooper over and have me summarily removed for going where I wasn't invited. But if I were an elected official, I would be dubious toward every single report issued by the bureaucrat. I would assume that whatever the government agent said, it was designed to hurt the good folks and protect the bad ones.

I've been to Richmond countless times, promised by my elected delegates or senators that a hearing will extend "equal time." Dear readers, it doesn't exist. After one delegate actually asked me to accompany him on a day of lobbying for a bill that would exempt on-farm sales from onerous regulations, I confessed to him at the end of the day that I was frustrated with the system. All day long we could scarcely get an audience with a legislative aide.

But behind my shoulder, whenever an industry lobbyist came along, the door to the inner sanctum swung open and the politician himself welcomed the guest with open arms. It's sick.

Just sick. But the answer is not lobbying reform. The answer is for people like you, who are reading this book, to get involved. For us to join forces and scream "Enough!" and take back our government so that it serves our communities and not just ensconced global empire builders.

All of this over a piece of bacon. Yes, it really does come down to that. Until we realize that these big issues graphically and viscerally affect your freedom to have pastured bacon for breakfast, we'll never get anywhere. And this is an issue that the environmentalists do not understand, the Chamber of Commerce folks don't understand, the organic certification folks don't understand, the military doesn't understand, and the greenies and foodies don't understand. How our culture views freedom of choice creates the climate for whether a local food system can exist. And whether you and I have an option to Wal-Mart fare. Let's bring home the bacon.

Chapter 6

Salmonella

I had just stepped into the house to grab a drink of water when the phone rang: "I'm letting you know you'll be getting a visit from USDA egg inspectors very soon. We just had a salmonella outbreak in our restaurant and they're saying it was your eggs. I don't think so, but what can I say? We're closed."

I tried to mentally process all this as it came gushing out from one of our best chef-owned restaurants. The owner was calling to give me a heads up before the egg police came.

Over the weekend, he had served an uncooked Hollandaise sauce over Eggs Benedict, from an open bowl that stayed on the counter in the kitchen for five hours. A couple of customers had developed bloody diarrhea and gone to the local hospital emergency room, where stool samples confirmed salmonella, the common thread being dining at this restaurant for Sunday brunch, and eating the Hollandaise sauce.

As the chef explained, this open-bowl sauce is a petri dish for bacteria. And he had all kinds of fresh seafood, fresh meats, fresh garden produce with dirt dangling from the roots. And, as it turned out, the kitchen did in fact have, shall we say, some cleanliness issues. The restaurant closed for a week to do a complete white glove.

When the health department regional administrator arrived, she saw our eggs in the cooler and immediately determined that they were the culprit in the outbreak. She put our eggs on the report as the cause of the contamination and ordered them destroyed.

I immediately went to the field here at the farm and collected both fecal samples and eggs, and sent them to an accredited independent lab for a salmonella analysis. We paid $200 to get the four samples analyzed and waited an excruciating two weeks for the report. Sure enough, the USDA inspectors showed up here at the farm a couple of days later. They wanted to see our eggs in the walk-in cooler.

I hauled some out and they set up their lab—about a million candlepower light, similar to the one used by the oral surgeon when you go in for surgery. It was bright. Really bright. According to government protocol, a sample is 100 eggs. To be legal for sale, only one egg per 100 may have what is called "adhering dirt." That means anything that can be physically scraped off. It could be egg yolk from a broken egg, manure, a piece of straw from the nest box, a feather, or even a piece of cardboard from the egg carton. Anything that can be scraped off.

These two inspectors took out 100 eggs and began their analysis. It would have made a stand-up comedy routine—believe me, Laurel and Hardy had nothing on these two—if it hadn't been so serious. They picked up each egg and scrutinized it upside down and everything but inside out. Since we use non-hybrid brown egg layers, our eggs exhibit more genetic diversity than commercial eggs. We have eggs with spots, ridges, little calcium deposits, dark shells, light shells, elongated ovals, and shortened ovals.

These inspectors looked at countless eggs, trying to decide if that was a bump, a calcium deposit, a piece of dirt, a bit of pigment on a sea of paleness. They spent a couple of hours examining those 100 eggs under the spot light. Wonder they didn't get sun burnt under the intense lumens of that light. Anyway, when they were all done, they determined that we had two eggs out of 100 with adhering dirt, and therefore these eggs were illegal to sell. They were, in fact, inedible, which is the operative term for eggs that do not pass. Inedible eggs are illegal to sell. We were, in fact, criminals.

To help set the stage, keep in mind that we do not like to wash eggs if they are clean. When a hen lays an egg, she coats it with a shiny film that acts as a bacterial protection. It also reduces evaporation from the porous shell. If the egg is clean, it's clean and we don't wash it because we want to leave that protective film on it if at all possible. We have customers who ask for unwashed eggs, knowing that the film will be intact.

We process the eggs by hand, not machine. When we gather eggs, we place them in a wire basket, bring them to the egg room, and take out the clean ones from the dirties. When we're finished packing all the cleans, we have a basket of dirties that we place in a bucket of lukewarm soapy water and sponge off each egg. We put the eggs in plastic flats with drain holes in the bottom of each molded depression. Over night they drain and dry down, and those are the first eggs we pack the next day before we start the cleans. And then we repeat the whole cycle.

These inspectors, interestingly, never candled the eggs. Candling is a procedure that allows you to see inside the egg by using an intensive light and shining it through the egg in a dark room. It illuminates the inside enough to see how big the air space is, how runny the albumen (white) is and if the egg contains any blood spots or other abnormalities. If you make a claim like "Grade A" then the eggs must be candled. If you make no claim, the eggs do not have to be candled. We make no claims.

As I looked at the stash of 1,200 dozen eggs that had just been deemed inedible, I asked the inspectors what we should do. They were actually fairly cooperative, and suggested that we just go through them, clean them better, and then be ready for a re-exam in a couple of days. The inspectors left and we hauled all the eggs out and began going through them. A fleck of straw here, a piece of our washing sponge there, once in awhile a tiny speck of dirt. We went through them all. Although I've not done a scientific analysis, having looked at many cartons of eggs in the supermarket, I am convinced that were this rule strictly applied, no egg producer in the country could comply.

Two days later the inspectors returned, looked at the eggs—with no bright lights. Just a cursory glance, really, and deemed them okay. Suddenly the eggs were edible again. Amazing, to go from inedible to edible—it's almost like alchemy, turning straw into gold. Or transmutation, or something. I think these inspectors really missed their calling: they should have been magicians.

The important thing to understand here is that nobody checked the inside of the egg. When a carton of eggs says "Grade A Large" and has that identifying USDA shield on it, nothing about that inspection checks the inside of the egg. Nothing checks for the things that can hurt people. The inspection only deals with visually observable qualities: exterior shell and interior air space, thickness and blood spots. Blood spots don't hurt anyone; they are just yucky to see when you're eating an egg. It's more a psychological aversion than any real potential physical harm.

The point is that anything that could make you sick is never part of the grading system. People put a lot of credence in a system that excludes the most important part. This is typical of the inspection system: it checks for the least important things and doesn't check for the most important things. The grading system developed as a cosmetic check, not a safety check. Candling can definitely show interior deterioration if the egg is very old, but even an old egg can't hurt you. It might not taste very good, and certainly won't whip into a nice meringue, but it won't hurt you.

Finally our lab analysis arrived in the mail. I tore it open and was elated to find no traces of salmonella even in the manure. We could eat our chicken manure safely! Now I had confirmation that the salmonella outbreak was not our eggs, and eagerly waited to hear from the health department's findings. I waited. And waited.

After another week, I couldn't stand it, so I called the health department lady. Without telling her about our test results, I asked her: "What did your cultures show?"

"What cultures?" she asked.

"Your cultures of our eggs that you said had salmonella. I'm curious what you found."

"We didn't culture anything."

"Why?" I asked, incredulous.

"Because cultures don't mean anything. They just measure a point in time, and that has no real bearing on a pathogen outbreak. It may have been gone by the time we took a sample. Culturing the sauce or your eggs would have been meaningless."

I paused. This was definitely not the answer I was expecting, and I had to process this before going on. "Well, if you didn't culture anything, how do you know it was our eggs then? After all, you did put our eggs in the report."

"When I saw unwashed eggs, I knew that was the culprit because this strain has been found on unwashed eggs."

"Only on unwashed eggs, or other things too?"

"Other things, yes, but also on unwashed eggs."

"Then how do you know it was our eggs? Maybe it was something else."

"This was my determination."

"The owner-chef told me it could have been any number of things."

"Unwashed eggs are inedible."

"Beg your pardon?"

"Look, consumers expect that their eggs are washed. In my opinion, an unwashed egg is inedible."

"Have you ever seen commercial egg farms washing their eggs?" When we first ramped our egg production up to several hundred dozen a week, we would take them to a nearby egg farm. This was a typical battery caged setup with 50,000 birds in one house adjacent to the packing house. The chickens would lay the eggs on the floor of their wire cage (9 birds in a 16 inch X 22 inch cage) which slanted toward a rubber belt conveyor. The conveyor delivered the eggs straight into the processing facility. No nest, no straw, no privacy.

A gentle amalgamation technique pushed the eggs onto little rollers that carried them first into a hooded washing machine. A 100-gallon reservoir held the wash water and it sprayed over the eggs while little brushes wiped the eggs as they rolled along. Amazingly clever—and stinky. A clean batch of wash water contained soap and tons of chlorine. As the eggs came along, of course, some would break. Shortly, the water turned yellow. Of course, manure built up in the water as well.

Before long, what started as a nice pristine-looking heavily-chlorinated dish water became a putrid, smelly hot shower of yellow sulfur-smelling manure-laden gunk spraying over the eggs. And it stunk to high heavens. Remember, eggshells are porous. A tiny bit of that cleaning liquid permeates the egg.

With all this in mind, I asked the health department lady why this type of cleaning was better than just leaving a clean egg alone.

She refused to acknowledge that an egg could be clean otherwise, and refused to discuss the aforementioned fecal cleansing method. Although she admitted that the regulations did not require washing, she simply sniffed and said; "An unwashed egg is inedible."

Then I told her we did in fact culture the eggs, including the manure, and they were clean. It was like talking to a fence post: "An unwashed egg is inedible." End of discussion.

I hung up and immediately the phone rang, even before I could take my hand off the receiver. I answered and it was Allan

Nation, editor of <u>Stockman Grass Farmer</u> magazine, the world leader promoting grass-based agriculture. He and his wife, Carolyn, had just returned from a fact-finding trip to Europe, with a heavy emphasis on range poultry in France.

Without any prompting from me, he started the conversation with, "Hey, guess what? In France, you can't sell a washed egg. If it has to be washed, it must go into a pasteurized product in liquid form. Only unwashed eggs are legal to sell as shell eggs."

Talk about a serendipitous phone call. What are the chances of that? I began laughing, and told this story and how I had just hung up the phone with the restaurant inspector who had a vendetta against unwashed eggs. He thought it was pretty funny too.

Now to tie up some loose ends in the story. The USDA egg requirements do not require eggs to be washed. We asked the inspectors specifically, and they even gave me a copy of the requirements. Absolutely nothing about washing.

When I called the USDA supervisor to talk about how to proceed with this health inspector tyrant, he was very open and explained that I was another victim of the turf war between the two oversight agencies. The USDA inspects farmers, but has no jurisdiction over restaurants. The health department inspects restaurants. He said the health department thinks the USDA is too lax and lets farmers get by with murder. He said the USDA thinks the health department is overly tyrannical and capricious.

And that's the regulatory climate under which we work. It's not clearly defined. When the health department sent the USDA guys out here to the farm, they didn't trust the health department anyway. But the USDA inspectors felt obligated to find something or they would be accused of being prejudicial toward the farmer.

I wrote a scathing letter to the governor and our politicians describing the incident and asking for an apology from this slipshod health department lady. No responses. Nothing. But she got mad at us.

About a year later, our farm was featured in the Smithsonian Magazine and included in the piece were several interviews with some of our chefs. Of course, they spoke glowingly about the quality of our products, especially the pastured eggs. This lady got a copy of the article, and actually carrying it under her arm, went to the three restaurants interviewed in the article and threatened to close them down if they did not immediately discard our eggs.

What's a chef supposed to do? Our culture has been conditioned to believe the government bureaucrat who captures the headlines with: "Eggs served at local restaurant pose health risk. Ordered to destroy allegedly inedible eggs."

You liberals reading this, please, please, put yourself in the shoes of the average American and honestly ask yourself what you would think if you saw such a headline. You have to admit, every American except the most rank libertarian would assume that the bureaucrat told the truth. The chefs are absolutely paranoid about something like this happening. They are in a no-win situation. They can't afford to fight it in the media, because they know they can't win. And the bureaucrat knows the chef can't fight it either.

The only way to deal with this is to bow, throw the eggs in the dumpster out back, and hope to goodness such a tyrant has a change of heart—or perhaps a heart attack. After the targeted visit, the chefs called us back and have been using our eggs ever since. But it's always a cat and mouse game, and they are always glancing over their shoulder, wondering what infraction they'll be accused of next. Egg cases, for example.

We want to recycle our cardboard egg cases. Many times when we go to the restaurant, we simply remove the flats of eggs, stow them in the refrigerator, and have a perfectly pristine egg case left over. Why throw it away? It's illegal for us to re-use that box because the egg police call them a one-time use vessel. Their justification for this view, of course, is cross contamination.

But let's be reasonable. If the case has some broken egg in it, fine. But cardboard boxes are one of the most benign transport containers ever known. The eggs don't even touch the box—they nestle snugly into their cartons or flats. The cases just allow us to carry 30 dozen of them at a time. If we put the flats in plastic milk crates, we don't have to discard or wash off the milk crates. Nobody cares what kind of vessel we bring the eggs in unless it is an actual egg case—then that is a one-use vessel.

As a result of this policy, we bring them home and re-use them as long as they look good, but we don't send them to restaurants. What a waste. Such a requirement is immoral, in my opinion. We should encourage recycling, not discourage it.

The thing we must all understand is that government reports must be suspect. One of my overriding goals in writing this book is to impress on the hearts and minds of Americans that official pronouncements from government agents are the result of an agenda that is driven by interests who do not seek the truth.

For the life of me I can't figure out how liberals who don't believe anything the Pentagon says somehow believe that another department of the government, like the Environmental Protection Agency, speaks the truth. This is an example of schizophrenic reasoning. Every one of these departments has an agenda, and the bureaucrats are not interested in the truth; they are political creatures wanting to expand their power, position, and possessions. People don't become divine or righteous just because they receive a government check and have alphabet soup credentials behind their name.

And why conservatives who don't believe anything the EPA says would worship the ground the Pentagon walks on is the same kind of schizophrenia. But we polarize each other. If I say the EPA has an agenda and is out to destroy business, I'm howled down as a rabid anti-environmentalist. If I dare to question the policies promulgated by the Pentagon, then I'm a pacifist dove who doesn't deserve the freedoms purchased for me by the blood of ancestors.

I think it's a much more consistent and reasoned approach to distrust everything that comes from the government, and everything that comes from multi-national globalists, and everything that comes from the media and talk show hosts, and everything that comes from mega-charitable organizations. And why do I seem to have such a prejudice against the empire guys? Because power corrupts, and absolute power corrupts absolutely.

And these health inspection bureaucrats have almost absolute power in many ways. They rule fiefdoms and enjoy a complicit, duplicitous American populace that assumes all is well as long as the fridge is full of beer, the toilet flushes, the TV remote works, and the sofa holds them up. To be honest, I am far more frustrated with complacency than I am with unscrupulous bureaucrats. If this book makes us angry with unscrupulous bureaucrats, I hope our righteous indignation will stir us to cast off complacency, and that is the beginning of integrity and accountability.

Chapter 7

1,000,000 Mile Chicken

"I want to take your chicken to the DuPont Circle Farmer's Market," said Bev, a quintessential marketer and gourmet food innovator. We had been working together on and off for several years and he wanted to move pastured meat and poultry into the belly of the beast—Washington D.C.

The FreshFarm Market's flagship venue was DuPont Circle on Sunday mornings. I had zero interest in going there, not just because of the time but because I didn't want the hassle of a three-hour drive one way. Besides, our farmgate sales were entirely sufficient for success.

But Bev, with that excitement that naturally grows out of a deep soul-level yearning, insisted that it would be a wonderful venue. His dream was to take heat-and-eat artisanal foods into metropolitan areas. My response: "Let them find their kitchens." Obviously, we shared complementary visions.

Thus began a saga that taught me much about the real world. I learned that "illegal" encompassed more than courts of law, deputies, and police. It included market rules and perceptions too ingrained to change. But let's let the story unfold.

1.000.000 Mile Chicken

The market really wanted chicken. No vendors had chicken. Bev wanted to sell the entire Polyface portfolio: beef, poultry, pork, rabbit, and eggs. Beef wasn't allowed by the market master.

"Why?" I asked.

"We already have a beef vendor."

"But what's wrong with a little friendly competition?"

"Might not be enough business for you both to be viable."

"So when will you allow us to come with our beef?"

"When I say so."

"What kind of criteria is that? It seems like you'd have some idea of when the time would be right to add another beef vendor, like when some customers begin asking for ours—which they already are—or when the other guy's sales hit a certain number, or whatever. His product is grain based and ours is grass based, so it's not even close to the same product, except that it is beef. It just seems to me that it would be fair to give me an idea of what criteria you are using to exclude our beef from the market, especially since we're already coming with our chicken and eggs."

"Are you disagreeing with me?"

"Well, kind of. How can you not have a standard on which to base our exclusion?"

"If you disagree with me, you disrespect me."

"How can we have a discussion then?"

"You don't understand, this is a dictatorship."

Where do you go with that? Folks, I am not making this up. This ended up being one of the most stressful, emotionally and financially draining marketing venues in which I ever participated. It went along like this for the couple of years we were involved with the market until we quit. We hung in there for Bev and his dream. And he was far more gracious and charitable through it than I. I lose respect real fast for people who get up on their back legs and tell me they are dictators.

Since we were getting some of our beef and pork slaughtered under federal inspection, we were completely legal to sell it at the market. But the market master didn't want beef—in fact, denied us the freedom to sell it. What she really wanted was chicken.

One small problem—Washington D.C. is across the state line. That's interstate commerce. And our chicken was not federally inspected. Big problem.

Ever the optimist, Bev said, "Okay, let's just get the chickens processed under inspection and we'll be completely legal." He began searching for facilities that would do it, and finally found one in North Carolina—more than four hours away from our farm and more than 8 hours from Washington D.C. , where the market was.

He sent an apprentice up with a flatbed trailer towed behind the refrigerated truck to pick up the birds at our farm. To put this in perspective, Bev's center of operations were 100 miles south of us. The processing facility was 200 miles southeast of us. That meant he had to drive 100 miles north, get the birds, and then drive 200 miles southeast to the slaughter house.

The apprentice arrived an hour before dark. He drove out into the pasture and we caught and loaded the birds into the crates on the flatbed. He headed out a little after dark, around 10 p.m. Somewhere around the mid-point of the trip, at midnight, he felt a nudge and watched in horror as the flatbed trailer detached from the truck and headed toward the guardrail.

It banged against the guardrail, thumping to a stop. Many of the crates flew open and chickens went all over the side of the road. A state trooper showed up shortly and spent the next two hours corralling and catching chickens. The apprentice, of course, called Bev about 2 a.m. to report the incident and explain that the kind trooper, who had the time of his life catching chickens in the sideditch, let the apprentice ease the trailer to a welding shop where they fixed the hitch. He got back on the road in time to make the 8 a.m. rendezvous at the slaughterhouse.

The next day the birds were bagged and cut up with a bona-fide USDA inspected label—the shield of everything wholesome and good. For the first time in history, a Polyface bird was blessed with federal credentials and legal to be sold across state lines. What a monumental day!

That next Sunday, those birds, which Bev transported 200 miles back to his center of operations in southern Virginia and then north 200 miles to Washington D.C., were proudly displayed for sale at the DuPont Circle Farmers' Market. And people loved them. They flew off the table. Bev needed more chickens. Time for another flatbed trailer trip in the middle of the night. These birds had traveled 200 miles live and 300 miles dead just to be sold at a market 150 miles from where they were raised. We hauled them at night so the birds would be more comfortable.

This time all went uneventfully. The third trip of the season, I decided to take them down because I wanted to see the slaughter facility and I wanted to help Bev out with all this pavement pounding. We loaded the birds in crates on our pickup, stacked 8 high inside cattle racks, and I headed out at 3 a. m. to make my 8 a.m. appointment.

The facility was quite small, perhaps 30 ft. X 50 ft. located in a rural area. The 15-person crew, all Hispanics, arrived and began dispatching the birds. As I move through this narrative, please understand that I had been processing birds for decades, on our farm, without inspection. To say that I was appalled at what I saw would be the understatement of the year. First, they electrocuted the birds before slitting the jugular vein. That shuts down the autonomic nervous system. The heart does not continue to pump, to push the blood out. The birds were poorly bled—this makes black around the bones.

The scalding and picking were fine—basically the same as we do it. Then the birds went onto shackles that moved slowly around an evisceration room, which was a little concrete cubicle with a slow-moving line of chickens passing along at eye level. Each person did a certain cut and the birds passed by the federal inspector, who looked at the entrails and carcasses for signs of problems.

Of course, the inspector didn't know me from Adam and certainly didn't know that I would put his comments in a subsequent book. I chummed up to him and engaged him in small talk. He normally worked at a large industrial plant nearby, and he said that our worst chicken was better than the best chicken he ever saw there. That was an interesting little tidbit.

The thing I'll never forget, though, was that a large pan of livers was on a table right in front of him. In other words, the chickens were coming along slowly at eye level, and at his midsection was a table with a big pan. The worker right beside him removed the liver after the inspector looked at it.

Let's get the context. The gallbladder is imbedded in a fold of the liver. It's a bright green organ that makes bile, the most despicably-bitter liquid known to man. The old saying "bitter as gall" undoubtedly came from a butcher who had accidentally popped a gallbladder and tasted the splatters. Anyone who has eviscerated a lot of chickens has a gallbladder experience. It just happens.

That's why my son Daniel tells folks who come to learn how to process chickens that his two rules are:

1. Keep your mouth closed.

2. If you feel something on your lip, don't lick it.

Whenever we accidentally break the gallbladder and the green juice splatters over the liver, we discard the entire liver. Clean-up is well-nigh impossible. Safer to just throw the liver away.

Imagine my horror when, standing next to the inspector, this worker was just ripping the livers out and breaking numerous gallbladders, which burst into the pan of livers. Here in front of the inspector was a large pan of livers floating in green juice. It was one of the most disgusting things I have ever seen. Now get ready for the punch line. These inspectors live by temperature. They worship thermometers. I don't know how any of them can explain human survival until the thermometer was invented.

Periodically, he would dip his little pocket thermometer into the liver pan to check the temperature. Sometimes he would say, "More ice," and other times he wouldn't say anything, presumably when he was satisfied that the temperature was okay. He never said anything about the green gallbladder juice in the livers. The entire chill water in that pan was dark, dark green. That didn't bother him at all. But boy that temperature had better be right. Sure brought to my mind the parable of straining at gnats and swallowing camels. I wouldn't want to swallow anything from that chill pan.

The whole evisceration procedure was sloppy. In our processing, we have one quality control (QC) person per pre-QC person in what we call the disassembly line. In other words, if we have an executioner, a scalder-dunker operator, and two eviscerators at the front end, we have 4 QC folks at the other end.

At this federal facility, one QC person handled the output of a dozen workers. The birds went into the chill tanks with stray feathers still affixed. I was not satisfied at all that the procedure was clean. But that apparently is okay as long as the temperature goddesses are satisfied.

Interestingly, the report for discards and blemishes showed that we had more problems than an industrial outfit. Why? Because here in this little facility, the inspector actually had time to look at the birds. They were coming by him at about one every twenty seconds. In the big plants, they go by at a fraction of a second apiece. That is why the government paperwork, which is dutifully picked up by Consumer Reports and other consumer watchdog groups, shows that small producers have more blemishes than large producers. It has nothing to do with actual fact; it's just a factor of speed and accuracy. A paper-manufactured discrepancy that contains no reality.

And yet Americans read these reports as if they are the gospel. The reports are merely an expression of what the system deems important—or what the system sees. In other words, the reports you choose to believe are a matter of faith. If the infraction was not seen, it did not happen. And if the rules of the game can be manipulated such that the referees cannot see the infractions, then no one is the wiser.

I walked out to where the processing water went. It just went out into a little wooded area and a quarter acre dead zone around the end of the discharge pipe. Federal inspection had no jurisdiction over the discharge water, I was told. That was the health department's problem. The health department didn't care about this because it wasn't human sewage. The result? Contamination and ecological disaster. Big lesson: just because an outfit is small does not make it clean or environmentally friendly. It probably does mean that by itself, it can only inflict a modest amount of damage. To that extent, small outfits have a little more saving grace than large ones, simply because they can only do a minimal amount of damage.

The owner of the facility, a hardworking entrepreneurial type, told me he could never stay ahead of the violations. It might be a tiny one-inch tear in a screen door. It might be a missed thermometer check. He said the bureaucrats never wanted to sign off on anything and just seemed to exult in badgering. A month later, he went out of business. Big lesson: asking for inspection and trying to conform to the system do not assure success. Often it just compounds the problems. Both Bev and the owner told me that the inspectors had conspired to run this little on-farm rural facility out of business.

In the inspection world, it was considered an armpit, a backwater. Right behind the inspector's station is a red button. At any time, the inspector can touch that button and stop the entire plant. That's how much power he wields. Wielding that power over a plant of 15 employees doesn't give the same rush compared to a plant of 1,000 employees. Something in the heart of man yearns to swagger around a bigger outfit, and inspectors are no different. They'd rather be with their friends in the prestigious surroundings of a huge complex. Anyway, the plant went defunct.

Now what to do? Bev searched for another federal inspected poultry processing facility, and he found one in Pennsylvania—200 miles north instead of 200 miles south of us. He came by and picked up birds and took them up to that facility. To our great dismay, nearly 20 percent of these beautiful birds were discarded—thrown away—because the inspector said they had a lung problem. He said they turned their heads into the wind on the interstate and the air affected their lungs.

Now folks, I may not be the brightest bulb on the rack, but I do know a few things. We transported half a dozen loads in the identical crates from the identical field on identical roads for an identical distance the identical summer. Not a single one had this problem. Ever.

All of a sudden, for some inexplicable reason, this batch of birds happened to look into the wind instead of away from it, and were judged inedible. This can't be a problem because the industry transports birds up and down the interstate all the time and I know the big guys wouldn't tolerate a 20 percent discard rate. But what do you do? When the inspector makes his judgment, the bird goes in the discard can. No negotiation, no ombudsman, no mediator. Unless you're McDonald's and hire attorneys from the political-governmental-industrial revolving door. The poor farmer is completely at the mercy of this bureaucrat. Imagine having that kind of autocratic power at your fingertips, to just be able to make those judgment calls without any oversight. It's downright scary.

I don't believe a single thing was wrong with those birds. Perhaps we weren't the only ones that happened to, because a month later, that place went out of business too. Hmmm. I wonder why. But the owners of the plant can't buck the inspectors. If they irritate the inspectors, the inspectors have the power to make life unbearable. And of course the little outfits simply don't have the resources to fight back. Bev and I began feeling like bringing our chickens to a plant was the kiss of death. In just one short season we had been one of the last ones through the doors of two federal plants.

What to do? We were at the end of the season and had one more batch to go. Bev located another outfit, a large processor near Harrisonburg, only 60 miles away, that he cajoled into doing 400 broilers. This outfit normally processed cull hens, but the manager said they would work us in on the front end so ours would be the first birds through for the day. He was extremely accommodating, and I found him quite sympathetic to our needs. A genuinely gracious, sweet spirited manager.

We took 400 birds, and got back five in one piece. Want to read that again? Yes, that's right. We lost about 20 percent to the cull barrel again, but not because they looked into the wind riding up the interstate. This time it was because the automatic jugular cutter wasn't set just right and lots of them were not dead when they hit the scald water. Yes, you got it. They were scalded to death. When that happens, they don't bleed because their throat was never slit. The industry calls these cadavers. They are red when they come out of the picker because all their blood is still inside.

All of those went in the discard can. Of the remainder, every single one was beat up in the picking gauntlet. Dislocated legs, broken wings, torn breast skin. They were just torn to pieces. Only five came through unblemished. Bev and I looked at each other, "We can't afford to do this again," we said in unison.

The last two outfits we had used did not offer a complete service like the first little operation did. Although the first two facilities we had used were small, they did everything from slaughter to the final bag. The first one even cut them up and put them and packed them onto tray packs. Of course, we needed the birds in a final USDA stamped package in order to sell them at the DuPont market.

The third facility, however, was a huge industrial facility and it did not offer such a complete service. The only way this outfit packed birds was in waxed cardboard boxes with some ice on them. No individual packages or anything. That required us to take them to another facility that would bag and cut them up. And that facility had to be federally licensed too. We couldn't do this at home.

He located a facility just over the Virginia line in Tennessee, near Bristol. So the Pennsylvania birds traveled 200 miles from our farm north to be killed. Then they went 400 miles south to be packaged. Then Bev hauled them 300 miles back up to DuPont Circle to be sold 150 miles from our farm where they were raised.

The Harrisonburg birds went through the same transport distance. This simply was not working. By the time we split our losses, and Bev paid me, and he paid for all the transportation miles, even at this upscale metropolitan market he couldn't charge enough to cover the costs. Both of us lost big time.

That winter, during the off season, Bev concocted a plan. As a fellow conspirator, he was a great sidekick. He realized that the powers that be at the market never really checked the labels, so he decided to just take our on-farm processed chicken to sell, but display the leftover federally inspected birds.

When someone wanted a chicken, he would just reach behind him to a cooler and pull out one of our on-farm processed birds, stick it quickly into a bag, and leave the display ones there. If anyone noticed his sleight of hand, he would just put on a theatrical, "Oh My!" and disclaim: "Oh, I must have picked up the wrong one by mistake." At the Virginia markets a couple of miles away from the D.C. market, our on-farm birds were completely legal. Amazing how those little red lines on a map can make food either safe or hazardous.

That's exactly what he did the next year. All year, we dressed the birds here and he kept a box of display birds with the federal shield on them. The ruse worked beautifully. It was illegal as sin, but no one was the wiser. When the market masters found out, what could they say? All the customers were enjoying the game and loved the chickens—they could actually tell the difference between the federal inspected birds and the on-farm birds. The on-farm birds were cleaner and tasted better.

By the end of the season, we both split our sides whenever we looked at those poor federal inspected birds in their box, the 1,000,000 mile birds, and agreed that some day, when the danger of being fined for this infraction that threatened the sanctimonious food safety net in America was long past, we would come clean with our great chicken caper. Here we were, passionate proponents of local food, which includes shortening the distance between producer and plate, and we were putting more miles on birds than the industrial system we preached against. But they were just the display birds.

This experience, more than any other, made me fearful of joining the system. Bev said the inspector at the small plant made it clear to him that the inspectors did not like going out there and were planning to put it out of business. It did not have recreation rooms for inspectors, the camaraderie and prestige that come with being in charge of a big operation. I guess these inspectors are kind of like church pastors after all. Ever notice how pastors who move to other ministries always get "called" to a bigger church? Why don't any of them ever get called to more humble circumstances? Amazing how that "calling" always seems to go one direction.

Oh, but I digress. So many inconsistencies to take broadsides at, no? Later, when Bev started his own slaughterhouse, the inspectors wouldn't come because they said he was too slow. I don't see anything in the law that says you have to be fast enough to earn an inspector. If you comply with their standards, they have to be there. But not when you're a small, struggling dreamer. They can just kick you around like a mangy mongrel.

I asked a federal inspector once, who complained to me that he routinely went into clean places and filthy places—all inspected, mind you – about the possibility of empirical testing. I asked him why they couldn't just put thresholds on what's allowable. After all, the Food and Drug Administration clearly articulates how much rat manure can be in granola, so why can't the Food Safety and Inspection Service just articulate so many parts per million of *salmonella* or whatever? We could just do a swab test and if it exceeds the allowable limit, an inspector shows up and we deal with the problem.

He agreed that such a plan would be completely efficacious, but immediately turned negative with this response, "Then I'd be out of a job." Isn't that just like life? Here you have a simple, elegant, workable solution and it all boils down to my job. Never mind that anyone willing to work can always find work. Is the goal of the system to insure job security for needless inspectors? An objective test really would put everything on an equal footing. If I can eviscerate a chicken in the kitchen sink as cleanly as the threshold, who cares if it was done in the kitchen sink? Aren't we really interested in clean? Or are we really interested in protecting jobs, keeping our turf, and being the Grand Pooh-bah?

This reluctance to actually help food be clean is what keeps me from ever wanting to join the system. I've talked to numerous small operators who have been jerked around and jerked around by these inspectors. One lamented to me, "On Monday, they want the room red. Next Monday, it has to be green. And the following Monday, it must be yellow." Average people don't have a clue how much of the regulations are subjective.

When Bev was writing his HACCP (Hazard Analysis and Critical Control Point) plan, he pulled the template right off the FSIS website. Why not use the boilerplate from the official government source, plug in the appropriate numbers to customize it to his operation, and not try to re-invent the wheel? That approach sounded reasonable to me.

But when the inspectors came to look at it, they said it was all wrong. Incredulous, he told them he got it off their website. They said, "Oh, that doesn't mean anything." Folks, I can't make this stuff up. Truth is indeed stranger than fiction. I ask you, how would you like to be a guy who's leveraged every penny to his name for this dream, only to have bureaucrats like this just come in and jerk you around. When he asked them if they would help him, they said he just had to write it and they would tell him if it was okay. Just an arrogant, uncaring, malicious bunch of bureaucrats. A year and lots of emotional agony, and lots and lots of salary later, he had something they said was okay. It's devious enough to make even the most optimistic yell "Conspiracy!"

They give permission for something one week and then retract it the next week. Just like they told us when we had our custom beef run-in, "We just decided to interpret it differently." And none of this has anything to do with food safety.

The government encourages folks to go out on a 70 degree November day, gut-shoot a deer that may have Creutzfeld Jakob disease (mad cow equivalent), drag it a mile through the squirrel dung, sticks, and rocks, display it prominently on the hood of the Blazer to parade the trophy around town in the afternoon sun, bring it home and string it up in the backyard tree under where birds roost, wait a week, then skin it out, cut it up, and feed it to their children. That's safe. Give me a break.

Is it impossible to ask the system to self-regulate? When Upton Sinclair wrote The Jungle in 1906 exposing the corruption and filth in Chicago's meat packing industry, beef consumption dropped nearly 50 percent in six months. How long would the industry have sustained that before making changes? Not long.

Very shortly private certifying agencies would have sprung up, like AAA for beef. You join, we inspect. But instead, Americans clamored for government relief and the Food Safety and Inspection Service was born. Today, the corruption is worse, filth more deadly, and the consumer less protected. Meanwhile, the real alternatives can never see the light of day because the hoops fledgling wanna-bes have to jump through are too high to reach. And too capricious. It's like running a race toward a moving finish line. Or playing a game in which the referees change the rules every 5 minutes.

We have chefs begging for pig's feet. No problem, except that the inspectors have recently deemed these heritage delicacies inedible unless the hog is scalded. If the hog is skinned, the feet must go out in the offal. It must be scalded and scraped, with the skin left on, in order for the pig's feet to be kept out of the rendering barrels. No amount of arguing can do any good. Since our processor doesn't scald, our chefs are out of luck. Those feet could easily be scalded—even in the restaurant sink.

My point is that it's illegal for me to even get those feet back. I cannot walk into the plant and acquire those feet—it's illegal. I can't give them away. I can't cook and scald them myself, to use for myself, to eat myself. They have to go into the waste stream. These rules are unconscionable, and play into the hands of the large players that process enough hogs to afford the $100,000 machines to scald efficiently.

The 1,000,000-mile chicken illustrated the artificial barriers keeping folks from acquiring affordable, cleaner, local food. None of this had anything to do with safety; it was all bureaucratic mumbo-jumbo, paperwork, and foolishness. Probably the worst part was that the market masters were and still are unwilling to use their political pull and savvy to acquaint people with these issues.

They were too busy making sure that all of our paperwork was filled out and we weren't competing with another vender. Since leaving that farmers' market, we've never gone back. The politics of the market, the rules, the heavy-handed bureaucracy all worked together to make it an emotionally and economically draining experience. Amazingly, I've learned that most people involved with the market thought we made a ton of money there. That's one of the biggest jokes around Polyface farm.

Whenever we need a good laugh, we just look at each other and say: "Remember all the money we made at DuPont Circle Farmers' Market?" That's a real lifter.

The Present

Chapter 8

Organic Certification

T he letter arrived in the regular mail one day. "You are in violation of the organic certification law and we will begin court action if this infraction persists." The threat from the Virginia Association for Biological Farming (VABF) was like being sued by my best friend.

My relationship with VABF was long and precious. Founded just a few years before my Dad passed away in 1988, it was the support group for non-chemical farmers. Nearly every state has one of these organizations:

· Carolina Farm Stewardship Association
· Ohio Ecological Food and Farming Association
· Pennsylvania Association for Sustainable Agriculture
· Georgia Organics
· Northeast Organic Farming Association

These are just a tiny sample of the many that exist. Most came into being in the 1980s to give those of us in the clean farming movement helpful collaboration for information and inputs when this was much harder to find than it is today. Several of us could get together and order a tractor trailer load of greensand, for example. Tapping into other farmers' experience, the collective information pool was extremely helpful.

As the movement grew, of course, the infrastructure developed along with it. Today information and ecologically-sound inputs are far more available than they were in the late 1970s. Farmers like us joined together with like-minded farmers in these early affiliations as an alternative to the extension service.

In those days, finding an environmentally-sensitive professor at a land grant university was about as rare as finding a vegetarian at a barbeque. They just didn't exist. Today, this earth-friendly movement has penetrated the extension system and friends do exist, struggling against great odds, I might add.

One of the worst gag cases coming out of that era was when Roger Wentling, a columnist for <u>Stockman Grass Farmer</u> magazine and a fulltime agent with the Soil Conservation Service, began aggressively promoting grass dairying in Somerset County, Pennsylvania. An absolute evangelist for the benefits of grazing dairy cows rather than confinement feeding them, his success in that area was virtually unprecedented. He would go out and build fence for dairy farmers.

His enthusiasm literally converted the entire area into what is still today a bastion of grass-based dairying. His success, though, put him crossways to the local feed mills in the area, which saw their grain sales plummet. The feed mill operators complained to Wentling's superiors and the agency slapped him with a gag order. He could not speak about grass dairying any longer. <u>Stockman</u> magazine ran a front-cover picture of Wentling kneeling in a beautiful field of clover, handcuffed, with duct tape over his mouth. It showed the power of the system to protect its turf when truth begins to invade its environs. So much for government openness to new ideas.

Our family became active in the VABF right when it began. In fact, our son Daniel was only three months old in 1981 when we went to a VABF field day and on the way heard Raymond and Dorothy Moore introduce the term homeschooling for the first time on James Dobson's Focus on the Family radio broadcast. We immediately ordered their books and that provided the impetus to homeschool our children.

The year I left the local newspaper and returned to the farm fulltime—Sept. 24, 1982—the VABF monthly journal needed help. As with all volunteer organizations, the folks putting together the journal were beginning to wear out. I quickly volunteered and the journal committee decided to make me editor. Of course, nobody else wanted the job: no resume needed.

Without a computer in those days, I typed all the articles and physically cut and pasted them on the blueline layout sheets. Then I would take the paste-ups to a printer and someone else put the mailing labels on them. Unfortunately, a year later I caused a great schism in the group by publishing an investigative article about one of our dairy goat members being harassed by dairy inspectors. The expose was not what the anti-political VABF hierarchy wanted. As a showdown developed, I resigned rather than face the rack. I was editing the newsletter for free anyway.

A couple of years later, though, I was elected vice president and the following year, president. After being president, I remained on the board. The point here is that many of my dearest friends and associates in biological farming, endearing relationships, were forged during these formative days. I was able to tap into the wisdom of a handful of elders who had been building compost piles long before I came along. True visionaries. In fact, two of them were Plowboy interviews in <u>Mother Earth News</u> magazine. These were icons for me, and to be in their homes, rubbing shoulders with them, was truly an amazing experience for a young buck like me with more sap than brains.

I consider this group of A.P. Thomson of Golden Acres Apple Orchard fame (widow Scottie and son John are still operating the orchard, one of the oldest and best in the world), Ellie and Don Pruess (founders of the Bonanza restaurant franchise who escaped New York after being burgled twice in a month and began grass farming in Virginia), and Reid Putney (whose organic beef cattle operation was featured in the <u>Wall Street Journal</u> in the 1960s) as my mentors. These people became surrogate visionaries for me after my own Dad's untimely death in early 1988.

Then along came organic certification. Interestingly, none of these original founder-mentor-visionaries of VABF wanted anything to do with certification. They just wanted to continue helping each other in this great common cause toward environmentally-friendly farming. Many of the younger members, who were also my friends, were confident this would be the best thing that could ever happen to our clean food and farming movement. They said it would quintuple membership. During all of this hoopla, I threw the cold water. The rain cloud on the parade. I was the dampening spirit. It strained friendships and I felt increasing ostracism.

My take on it was that it would divide the ins from the outs. It would form cliques instead of collaboration. It would create infighting and drop our membership rather than stimulating it. And if the government was involved, I predicted that it would be a political process and the smallest farmers would be hurt.

Long before all this certification discussion, we used the word organic freely and liberally to describe our farm. We hadn't used any pesticides or herbicides in decades. We fertilized with our compost. We did buy grain from local growers that wasn't organic, but our pastured poultry received a new salad bar every day. In the big scheme of things, we were the quintessential organic farmer.

A Washington D.C. food group contacted us about being listed in their directory of clean food farms. We agreed to be listed and wrote up a little blurb about Polyface, using the "O" word. Foodie groups, natural food stores, food coops and others used the directory to stimulate connections between buyers and producers. It was a great little pamphlet type directory, perhaps a dozen pages in size.

Now back to the VABF's certification juggernaut. Many in the group wanted to be the official sanctioned certifying agency in the state. They believed the prestige was not only due our fledgling organization, but would establish us as the official voice for biological farming. Everyone pushing the certification agenda had desires that were as true blue, sincere, and pure as the wind driven snow. Many saw this as a real victory in official state recognition that we were a credible movement, a force to be reckoned with. After a lifetime of being pooh-poohed by the official powers, the thought of being wanted and needed was indeed a heady notion.

When I saw that my side was outvoted by a huge margin, I exercised the only option left. I made a motion that VABF become the certifying agency for one year. After one year, we would have to vote on it again to continue. This sunset clause, I thought, would give enough time to prove my concerns correct. I reasoned that within a year enough of my predictions would come true that a majority would not vote to continue, and it would be dead in the water. Then, I reasoned, the group could get back to its original intent, which was to be facilitators rather than policemen.

Of course, everyone was more than happy to vote yes to my motion, glad that the vocal dissenter had acquiesced and unity was restored. Those were difficult days of soul searching for me. I never want the reputation of being a negative person. I want to exude a can-do spirit, to be an encourager, not a discourager. I don't even write nasty letters against the World Trade Organization. I'd much rather create change by liberating the bottom rather than regulating the top.

The group promptly put me on the certifying board. We hired an inspector for the requisite third-party independent verification and began taking applications. I attended only one certification session before resigning. We had about ten applications to process and vote on, and as each came before the board, I was immediately struck by the partisanship not only on the part of the inspector, but also on the board.

If the applicant was a big operator, a mover and shaker, as they say, the board virtually rubber stamped everything. But if the applicant was small, and especially if the applicant was known as a little bit more radical than most, we questioned, scrutinized, put on stipulations or even denied. The discussion centered not at all around real practice, but degenerated to personality profiles and political prowess. It was not at all what I had predicted: it was far worse. I never went to another session.

At that point, I kind of dropped out of the VABF. Not an enemy by any means. But I just didn't have the stomach to throw myself, my time and energy, into a volunteer organization that I believed was really headed the wrong way.

A couple of years later, the letter beginning this chapter arrived, threatening me with legal action because someone spotted a leftover, dusty directory in a Washington D.C. health food store in which I had described our farm as organic. After all the hundreds of hours I had devoted to this organization, for them to threaten me with legal action because I had not ferreted out and destroyed every pre-certification use of the "O" word as being affiliated with our farm, was just like being sued by my best friend. The whole notion that I was legally liable for some dusty old directory written and published years before certification went into affect just didn't seem reasonable.

I made some calls to the VABF president and the state oversight agency. At all levels, they assured me that any statement I made about our farm was considered a label. And certification at its most fundamental level is a labeling law. I could not speak the word, in connection to Polyface, without being illegal.

Well, now, that was a fine kettle of fish. Because I had not gone through the certification process, I had no right to use the "O" word. It was taboo. I assured everyone that I would henceforth and forevermore cease and desist using such a deadly word, and I repented in sackcloth and ashes that it still appeared on a directory somewhere. They agreed not to prosecute, and in the negotiations, attacked me for not being organic.

"What do you mean, not organic?" I asked.

"You don't use organic grain."

"I can't find any locally."

"You don't use organic grain."

Was something wrong with my hearing? Seemed to be a refrain going on here. At one meeting a grain farmer in West Virginia stood up in the room and castigated me for not using his grain. He grew wheat. I didn't have wheat in my feed rations for poultry or pigs. No matter, he was incensed that I didn't buy his grain, even though it was not even in our state.

"I can't find any locally." If one could use a refrain, two could, I figured.

"You don't use organic grain."

I could tell this conversation would not proceed very much unless one of us came up with a different line. We weren't really covering new ground with all this repetition going on. "Is organic the only thing that matters?" I queried.

"You don't use organic grain."

Wow. This was getting difficult. I have found that fence posting is a favorite tactic of people who are long on doctrine and short on reason. Inability to think along with someone, to let the ideas move, is a hallmark of big mouths and small minds. I encounter this every time I attend a County Board of Supervisors Meeting to speak against property tax hikes to increase teachers' salaries.

My comments are short and consistent from year to year: "I was a product of the public school system. I had some teachers that I'd love to see get triple the pay they currently receive—they were that good. I had others that should have been fired long before I ever had them, and they are still warming seats in front of classrooms and wasting our time and money. But nobody has ever asked me which ones are which. And until someone does, the school system should not receive one more cent. If educators can't figure out how to reward good teachers and fire bad ones, then they don't deserve anyone's support."

By the time I'm done, the teachers, who always show up en masse to these things, are catcalling behind me and saying things like, "Why do you hate education? You unappreciative, small-minded person." I even had the county school superintendent corner me in the hallway after one of these hearings, "You have no right to even speak about the school budget because you don't have your kids in our system." Talk about small minds. I wonder if it ever occurred to him that I help pay his salary? I do pay property taxes.

Anyway, I decided to try a different tack with the organic grain guy: "If I have to truck this organic grain from 500 miles away, in the big environmental footprint picture, which is better? Grain I buy from the neighbor, who may have used non-composted chicken litter and a quart of Atrazine herbicide per acre, or the diesel fuel expended to get organic grain to me?"

"You don't use organic grain."

How do you deal with that? I decided I probably couldn't penetrate his brain, so I politely terminated the conversation to pursue other more intelligent things. The certification idea is a pass-fail system, and as such always moves toward a minimalist approach. And because it is a non-comprehensive term, it does not include many of the more important variables in a socially and environmentally responsible food system.

For example, we buy grain from neighbors partly to make our money circulate in the community. If we export it out of the area, my neighbor becomes less profitable and may do something like pave over his land for a strip mall. Does that part of the equation carry no importance?

Since certification has matured over the past decade and a half, of course, it has done precisely what I predicted would happen. It has locked out small producers. It has pitted the ins against the outs. It has added another whole layer of food police. And it has aided and abetted the empirization of organics.

An outfit with 10,000 chickens crammed in a house who never see the light of day or set foot on pasture can certify them as organic. But mine that receive a fresh salad bar of pasture every day and really do taste different than industrial fare can't be certified because I'd rather use my neighbor's corn than ship it in from out of state. One season we decided to raise a group of bona fide organic chickens. We bought organic certified grain.

We had a batch of 1,200 broilers, and we raised 400 of them on the organic grain and 800 of them on our local GMO-free grain (GMO—genetically modified organism). The organic birds were inferior in every way: higher mortality as chicks, poor growth, poor feed conversion, and sickly. We finally discovered the reason: the corn was full of fodder. In other words, the harvesting combine was not adjusted correctly and the grain was full of pieces of the plant rather than just kernels of corn. It was a sloppy harvest job. And the fodder has virtually no nutritional value. The birds were starving for nutrition while eating.

Organic certification does not address sloppy harvesting techniques. It does not address worker wages. Look at the variables in carrot production:

· Soil fertility can be with on-farm generated compost using a symbiotic animal component or it can be with compost using manure from an industrial farm, or it can be from a bag of organic compost generated miles away, or it might not be compost at all, but rather chelated high tech minerals and humic acid inputs, or the farmer might not use any soil amendments and just rely on foliar fertilizer, or soluble nutrients can be applied through irrigation water. Are you out of breath yet? We're just getting started.

· Seed can be open pollinated, hybrid, heirloom, grown with chemical fertilizer, or compost (see all the variables above), saved from last year's crop, purchased from industrial seed companies from the local horticulture store, through the mail, on the internet, delivered by UPS privately, or through the US Postal System.

· Weed control can be through hand weeding using illegal aliens, children, apprentices, neighborhood employees or a combination of all the above; or weeding can be done with mulching, or landscape fabric, or plastic, or cultivating with horses, by hand, or tractors. Weeds may be burned away with propane burners or wiped with biological herbicides.

- Irrigation can be overhead, subterranean, hand sprinkled, from wells, ponds, public rivers, on-site collected rainwater. The water can be motion activated through a free-form, applied with or without purification, with or without soluble nutrients added. Or plants can be mulched deep enough to retain moisture and minimize irrigation. Mulch can be all sorts of things, including newspapers, carpet remnants, old hay, leaves, straw, lawn clippings, landscape fabric.
- Harvesting can be done mechanically or by hand, with all sorts of labor (see above).
- Storage and cool-down can be an assortment of refrigeration, wicking burlap bag coolers, solar-operated coolers, bio-diesel generator-operated coolers.
- Marketing can be local, farmers' markets, Community Supported Agriculture, designated drop points, on-farm sales, restaurants, alternative food stores, conventional food stores, internet sales, bar-coded or not bar-coded, employee delivery drivers with or without health insurance, legal or illegal, subcontractors, or teamster members.

And I'm sure this isn't nearly all the variables. Now do you see my point that organic is not a comprehensive term? What the certification process has done is automatically shut down the inquisitiveness into all these variables, creating hardening of the categories. The bad part is that when a person says they are organic, the conversation ends because everyone assumes they know what that means. Clearly, a whole lot of questionable practices can happen under the auspices of organic. And this little list was just for carrots. When you go to animals, the variables compound.

The current lawsuits against the organic dairy industry attest to the divisions that the certification program has created within the movement. I've always said that if you want to certify something, certify the farmers' bookshelf and magazine rack. This movement has always been about a worldview, a value system. It is lived out from a deep inner conviction, not a codified system of dos and don'ts. If I'm feeding my mind and soul with the right stuff, my heart and hands will probably be in the right place too.

As it is, a 3,000 cow confinement dairy in the middle of a desert can be certified organic, but my family cow on pasture can't be unless I spend days filling out forms, pay money to an agency, and submit to bureaucrats tromping over the farm. And the pasture must have been chemical free for three years. This whole system has become extremely limiting.

One example. Just north of us some dairy farmers have gone into the certification program. One of them especially is doing a good job as a grass-based dairy and he would like to expand. A neighboring farm came up for rent. Lots of great grass—but he couldn't use it because it wasn't certified. Instead, he's buying exorbitantly expensive certified grain to supplement his grass in order to produce enough milk. The milk would be far better if the cows were grazing artificially-fertilized grass than eating organic grain because herbivores do not naturally eat grain. But feeding the urea-fertilized grass would be illegal. And that's a shame. I'm not advocating chemical fertilization. All I'm suggesting is that criminalizing chemically fertilized grass in favor of unnaturally-fed corn is not a rational tradeoff.

I'll take the integrity and transparency of local any day over government-certified organic. For the life of me I can't figure out how a whole movement predicated on the notion that the government food system was corrupt beyond salvage allowed itself to get hijacked by the government as the repository of high quality food protocols. This is schizophrenic reasoning at its worst.

I know how it happened. The organic movement, for the most part, grew out of a government-friendly political climate. I wish I had a nickel for every time I spoke at a conference and everyone assumed that since I was an environmentally-friendly farmer I must be:

·   In favor of the National Education Association's agenda, including being anti-voucher and anti-homeschooling.
·   In favor of more wilderness areas and expansion of public lands.
·   Opposed to private property rights.

- Opposed to tree cutting; in favor of tree hugging.
- In favor of labor unions.
- In favor of higher taxes to cure all our ills.
- Opposed to any military anytime anywhere for anything.
- Opposed to the individual right to keep and bear arms.
- Opposed to privatizing social security.
- In favor of abortion.
- In favor of mandatory seat belt ordinances and motorcycle helmet requirements.
- Worshipper of the creation rather than the Creator.
- Anti-Semitic.
- Pro-Islamic.

As a Christian libertarian environmentalist capitalist, I'm all over the board politically. Just so you'll know how weird I am and at the risk of offending absolutely everyone, I'll list off a few things:

- Against prisons. They do not reform and they are incredibly expensive. Re-instate the whipping post and/or put on surveillance ankle bracelets and require all robbers to make restitution. Rapists should be executed. I'm hard on criminals, but do not adhere to the "lock them up and throw away the key" thinking. Quick and appropriate punishment, then a second chance.
- For legalized drugs. All of them. The philosophical justification for the government to control cocaine is the same rationale to deny raw milk and homemade pound cakes. Why should the government be the one to determine what I can or cannot ingest?
- Privatized education. Shut down the public school system and privatize the whole thing with vouchers. Deep down, I don't think it's the government's responsibility to educate anyone, but that's probably too weird.
- Flat tax. Abolish the Infernal (sic.) Revenue Service and go to a straight consumption tax. Forget all the machinations of the tax code. Especially abolish the death tax—kill it. Death should not be a taxable act.

·   Mileage tax. Abolish the gas tax and institute a mileage tax. As fuel efficiency increases and alternative and non-taxed fuels proliferate, roads should be paid for by the square footage occupied per year. A gas guzzler getting 15 miles per gallon doesn't cost any more road maintenance than a hybrid getting 50 miles per gallon, but they occupy the same square footage of asphalt.

·   Shut down ALL foreign military bases. Quit building empires and overthrowing and propping up puppets. Quit meddling in other peoples' affairs. Have a defensive posture second to none and be a good neighbor, not a meddling neighbor. A sign hanging in one of the restaurants I supplied eggs to in my teen years read: "A good monkey is a monkey that doesn't monkey with other monkeys' monkey."

·   Stop ALL foreign aid. Period. Even in disasters. If charitable organizations want to solicit for money, wonderful. But the government's money is my money and it should either be left in my pocket or spent at home. If taxes were reduced accordingly, we'd all have more money to give to the charity of our choice.

·   Phase out social security. Why should you be forced to take care of me?

·   Get the government out of health care. Completely. If I want to build a hospital that only admits bowlegged cowboys and dispenses snake oil at midnight in rooms blaring "Your Cheatin' Heart" then I should jolly well be able to do that. The free market is dynamic enough to care for any need.

·   Eliminate the Bureau of Alcohol, Tobacco, and Firearms. Let 'em roll. Had we never prohibited alcohol, the healthy alcohol fuels industry would have never collapsed and we wouldn't have become so dependent on foreign oil.

·   Eliminate ALL farm subsidies. Actually, it would be fine with me to shut down the USDA; never before has any agency been so successful at annihilating its own constituency.

- Eliminate all corporate welfare. No business gets any tax concessions, free stop lights, free water hookups, or guaranteed spousal jobs in the public school system.
- Eliminate logging on government land, and privatize all national forests and probably all the Bureau of Land Management Lands. These public lands are political footballs, offering under-valued timber to drive down prices so private landowners have no incentive to steward their own forests. If the recreational folks want an unmolested forest, let them pool their money and buy it. Then they can build their own roads and drive their off-road vehicles wherever they want. At least we won't tie up countless acres and hours in this incessant tug-of-war between loggers, environmentalists, and recreationists. If the environmentalists want unmolested acres, let them buy it and pay for fire suppression themselves.
- Abolish the minimum wage and child labor laws. The reason we've got roving teenage gangsters is because all the productive things they used to do to make them tired at night are now illegal. Let them work if they want to, and quit denying them the privilege.

I probably ought to stop there. I'm sure everyone thinks I'm a kook now, but that's just a smattering of ideas. I spend a lot of time walking through pretty green fields contemplating what a different world we could have. I certainly don't have all the answers, and some of these ideas I'm sure are wrong. But if we aren't willing to at least throw them out there, we can't grow and refine them.

After all this, organic certification seems like a tempest in a teapot, doesn't it? And that's just the point. Doesn't our society have more important issues to wrestle with than whether or not I can legally use the "O" word? I mean, really.

If all the effort—time, jet fuel, wining and dining—that went into making sure I couldn't use the "O" word had gone into lifting the kinds of marketing restrictions articulated in this book, the clean food movement would be light years ahead of where it is. If we could actually sell to our neighbors without bureaucratic

119

involvement, we environmentally-friendly farmers would spin circles around Wal-Mart. Wouldn't that have been a more noble goal than crisscrossing the country in jets lobbying to make sure I couldn't use the "O" word?

I'm not saying nothing good came of the effort; all I'm saying is that we must carefully pick the most efficacious battle. None of us can fight all the battles that are worth fighting. We must pick the ones with the best return. A locally viable food system, I suggest, is superior to a global organic outsourced empire food system.

I disagree with the folks who say that certification has increased the organic market. I will not yield these sales. I would suggest that had the same effort been put into breaking down local food barriers, we would be selling far more alternative food than we are, but it would be through cottage businesses, community canneries and processors, and neighbor-friendly retail venues. Just because some good may be attributed to the certification effort does not diminish the value created by the same effort expended on a different theme.

Chapter 9

Best Management Practices

I opened the newspaper one morning and in the local news section a prominent local farmer was pictured standing next to his brand new manure lagoon, a gift from the government via the Chesapeake Bay Foundation.

Lauding the farmer's interest in environmental stewardship, local Soil Conservation Service (SCS) (the old Natural Resources Conservation Service--NRCS) personnel gushed over this wonderful new technique to clean up the Bay. Kudos to the ecology-loving farmer. For the low, low cost of $50,000. And now the oysters will be healthy and rivers will run clean. The Chesapeake Bay Foundation, one of the nation's premier environmental organizations, applauded this new cost-share program as one of its crowning achievements.

The impetus, of course, for this program was the well documented manure load from Concentrated Animal Feeding Operations (CAFOs) finding its way into waterways. And in this case, the farms did not have to be very big to be significant polluters. Just for fun, let me run down some numbers to show how easy farm pollution is.

If we assume a 90-cow dairy, which is not very big these days, those cows are generating 1/3 of a pound of nitrogen per day out their back end. A conventional dairy will have a 4-acre field next to the lounging/feeding shed for the cows to lounge on. This sacrifice area, as it is often called, becomes worn down by the cows. The grass becomes clumpy and lots of dirt shows between the tufts.

In nature, nitrogen is metabolized by green material; in this case, forage. But the forage can only uptake a certain amount, based on climate and rainfall. In a brittle (low rainfall) environment, of course, the forage can't metabolize as much nitrogen because the climate simply can't grow that much forage. High rainfall areas can metabolize much more because water is not a limiting factor. In the Shenandoah Valley of Virginia, 120 pounds per acre per year is the maximum suggested amount of nitrogen that can be assimilated on pasture. That's assuming, of course, a nice, lush, pasture.

That means a 4-acre pasture, even if it were lush and thick, could metabolize 120 lbs. X 4, or 480 total pounds per year. Once the application exceeds that amount, it goes somewhere: either vaporizing off into the air, and we all know what that smells like, or leaching into the groundwater or surface runoff. In either case, the nitrogen leaves the farm in a harmful form.

Now let's go back to our 90-cow dairy and assume that these cows drop half of their excrement on this 4-acre loafing paddock. Their total nitrogen load is 1/3 lb. per day X 90 cows equals 30 pounds per day. Half would be 15 pounds per day. Remember that we computed our total capacity at 480 pounds per year. In how many days, even if it were lush, growthy pasture and not over-trodden sacrifice area, would we fill the total nitrogen capacity of the field? The answer: 480 divided by 15 equals 32 days. In just 32 days. And folks, this model is still being practiced on thousands and thousands of farms all across the fruited plain.

What about the other 333 days of the year? Where does it go? Yes, you guessed it. Right down the river. What about the half left inside the feeding shed? That gets scraped with a tractor-scraper, pushed into a manure spreader, and hauled out to other fields as fertilizer.

Manure can be a blessing or a curse, depending on the weather. Raw manure is highly unstable; the nutrients want to vaporize if they get dry or dissipate into water if they get wet. In either case, holding them on the farm takes real effort and planning. In winter cold or summer drought conditions, biological activity shuts down and the soil goes into a self-induced hibernation, or rest. Feeding the soil with nutrients at that time, whether organic or inorganic, results in the nutrients moving into unwanted areas, like down the river.

One slight exception to this rule is in northwestern Canada, where winter-applied manure freezes solid until spring thaw. The freezing stores it in suspension to prevent movement into the water. But this is the exception, not the rule. Environmental organizations have done a credible job of documenting the movement of this dormant-soil-applied manure. For environmentally friendly farmers, winter application of soil nutrients, and especially manure, is universally frowned upon, and for good reason.

Armed with the evidence that these loafing paddocks and winter manure applications were polluting surface and ground water, environmentalists pushed through legislation in Virginia and many other states to offer incentives to farmers to stop these polluting practices. And here's where the story takes its typical wrong turn once the government gets involved. The environmentalists, believing that government has the answers, turned to the land grant universities for advice on what practices would help stop this pollution.

In other words, the environmentalists won the ear of the politicians, who in typical fashion assumed that by throwing money at the problem, it would go away. With their hands stoked with money, the environmentalists went to the agriculture experts to determine how to spend the money. The land-grant researchers are in bed with industrial agriculture. Period.

Science is not objective. Scientists have agendas just like everyone else. Very seldom is pure research, in any discipline, actually performed. Not regarding genetic engineering. Not regarding human health and wellness. People who actually heal are called quacks by the American Medical Association. Political-agenda research is ubiquitous throughout the scientific community.

The agriculture scientists answered the environmentalists with what is called the official protocol for Best Management Practices (BMPs). These have now been written for many activities. Loggers now must follow BMPs for road building. The one time on our farm that we had a commercial logger cut timber in exchange for a road, he followed the BMP. We wanted little ponds when he crossed ravines. The BMP called for a culvert in the bottom of the ravine, and then of course road fill on top. But we wanted the culvert installed at the top of the fill in order to have a pond behind the fill.

Such a plan did not require one different shovelful of earthmoving and not one pound of additional steel culvert. But it would create small ponds to hold back sediment, create additional riparian zone diversity, offer flood control, and give wildlife additional drinking areas. But we could never get him to see the validity or superiority of this idea because it wasn't a BMP. You see, the people who wrote the BMPs for road building were road engineers. They were not ecologists or wildlife biologists or hydrologists. Road builders do not think about wildlife.

Sometimes the research agenda is just symptomatic of ignorance as the product of specialization. Sometimes it is subconscious, based on primal loyalties and pre-conceived notions. And sometimes it's just raw, consciously-determined political preferences, including, but not limited to, keeping research seed money flowing from philanthropists, who in the case of land grants are usually large industries.

The BMP for manure handling was, and still is, lagoons. These engineers, of course, come from a predisposition toward water-based sewage systems. (I've come to the conclusion that for a government engineer, the criteria for the best solutions are the ones that require the greatest amount of concrete, steel and lumber.) The design that wins the day is the one that's most consumptive, because that's the one that best greases the wheels of

Gross Domestic Product (GDP). And everything revolves around keeping GDP expanding. Just ask the Chamber of Commerce.

The plan to save the Bay, then, developed systematically through the political and scientific process. The most efficacious way to control this manure was by installing manure lagoons on farms in order to hold winter generated material long enough to spread it only on biologically active soils. Sounds reasonable enough, but the law of unintended consequences kicked in to thwart all the sincere intentions of the good-hearted people who pushed through the enabling legislation to save the Bay.

Oh, I'm getting ahead of the story. This is too good to go quickly. Allow me to savor it by unfolding it chronologically.

As I looked at that picture in the newspaper and realized how much taxpayer money was spent on this farmer, and how many more would soon be lining up at the public trough to receive their cost-share monies, righteous wrath welled up inside me. On our farm, we had purchased a used commercial wood chipper. When we worked in the woods, we stacked all the branches and then chipped them into a dump truck for carbonaceous bedding in the barn where we housed the cows during the winter.

For context, I will briefly describe our intensive housed animal model in the dormant season. For starters, we graze most of the year with daily-rotated pasture paddocks. This not only inherently spreads the manure from the grazing animals, but also minimizes the amount of time they are on stored feed. Non-grazing, of course, is the situation that necessitates concentration in a housing or tight lounging regimen, which in turn creates the pollution problem. The less we have to bring feed to the animals, the less we have to deal with having them cooped up in tight quarters. Rather than carrying hay out to the pastures in the winter and spreading it on the ground for the cows to eat, we bring them to the hay shed and feed them there.

This allows us to protect the fields from winter pugging damage and hold the manure in suspension with carbon to keep it from leaching into the groundwater.

In a nutshell, then, here is our winter feeding system. We have big open pole sheds—no walls. Lots of light—including big skylights-- and ventilation. The sheds have a central high core that stores hay, around which are awnings containing box feeders. We throw the hay into the boxes by hand to eliminate having to start tractors to deliver the hay to the animals. The lounge awnings, of course, catch the lion's share of the manure, and we put down junk hay, wood chips, straw, and sawdust as a carbon sponge to absorb all this excrement.

As this bedding pack builds higher, we lift the feeding boxes so the cows don't have to stand on their heads to eat. As we add additional bedding, we mix whole kernels of corn into it, which gradually ferments in the ever deepening anaerobic bedding. The cows tromp out the oxygen. The feeding shed smells like pine shavings and leaves and the animals have a warm, fermenting, clean lounge area. For a complete description and pictures of this system, please see my book Salad Bar Beef.

This bedding stays warm throughout the winter and grows natural antibiotics, exuding from the molds and fungi in the bedding. It often builds to nearly 3 feet deep. When grass begins to grow, of course, we turn the cows out to begin another grazing season. In order to convert these hundreds of tons of manure and carbon into aerobic compost, then, we turn in pigs, which seek the now fermented corn permeating the bedding pack. Like huge egg beaters, the pigs root through the bedding to eat the corn, and in doing so aerate the pack, converting it to aerobic compost. It heats and the composting digests the highly soluble and unstable nutrients, locking them into the skeletons of the dead microbes. This is the beauty of decomposition/regeneration.

In about 30-50 days, the compost is ready to spread on the fields, but it doesn't smell at all like manure. It has a wonderful sweet woodsy odor, and is stable against sun and rain. This beautifully elegant system has literally created Edenic pastures without machinery, concrete, or heavy energy use. Letting the pigs do the work completely changes the economics of large scale composting. The hay shed, then, becomes winter protective housing for the animals and their excrement, the fertilizer factory, and pig housing for a couple of months per year.

As I looked at that newspaper picture of the mammoth lagoon, my mind was also drawn to the numerous forestry conferences I had attended around the state and the ever-present hand-wringing over what to do about low quality woodlands and post-harvest slash. Virginia has lots of overgrown, abandoned fields that have come back in scrub and low quality pioneer species like Virginia pine. Professional foresters really don't have an answer for a market for these low quality woodlands. Nor do they have an answer for the millions of tons of slash left over from harvesting operations.

When I attended these seminars, I would take these foresters aside and encourage them to come out and see our composting operation. What I envisioned was replacing all the petroleum fertilizer used in the state with composting by utilizing all this excess wood. That would create a value for the wood that currently doesn't exist. Rather than exporting all our dollars for non-renewable petroleum, we could simply utilize solar-grown carbon in-state to fuel our fertilizer needs. Whole scavenging businesses could spring up, revitalizing rural economies and pumping more money into the forestry sector.

As I warmed up to the vision, these foresters' eyes started glazing over. Then they began glancing furtively around the room, hoping someone would come along to rescue them from this nutcase. On our farm, we figure the left-over slash value is worth about as much as the firewood from the same land area. But like most experts, the professional foresters have been taught to hate cattle farmers because cattle destroy trees and farmers don't give a hoot about good silvicultural practices. And of course cattlemen don't listen to foresters because foresters are off in la-la land thinking about 100-year cycles, so how could they know anything? And never the twain shall meet.

The U.S. has far too many dying trees. Please don't misread what I'm about to say. Nobody loves trees any more than I do. But young virulent trees pull far more carbon dioxide out of the air than old relics slowly decaying and then falling over. Trees are living things, which means they grow old and die. The proper use of our modern technology to suppress fire needs to be used in good silvicultural practice to cull the nonproductive trees and

128

encourage a new generation of young trees. We're simply not using enough wood in this country to maintain healthy forests.

I suggest that we displace all petroleum fertilizer by using wood for on-farm composting. That builds soil carbon which further sequesters atmospheric carbon, believed to be a key component of global warming. In the end, such a policy protects homes from wildfire, sequesters atmospheric carbon, and increases forest value, which in turn increases stewardship.

As I looked at this picture in the newspaper, my mind was running through all this reality. The fact is that on our farm we steward our manure beautifully with wood waste and composting. We don't want to put it in the water like a lagoon requires; we want to keep it out of the water. And the only special equipment we need is a chipper. Every farm already has some sort of dump cart or silage wagon or dump truck. My goodness, I reasoned, if the government wants to throw $50,000 at a farm to install an efficacious manure management system, why not purchase a chipper and hire the son or daughter that wants to stay on the farm but can't because the farm can't afford another salary? Put the child in charge of the fertilizer program via nutrient cycling. That answers the multi-generational farm component too.

Build the compost with pigs, thereby adding another enterprise and its valuable income stream. Make that eyesore 20-acre overgrown field a valuable biomass collector. The problem with lagoons is that they have a short life expectancy. The slurry pump is a high maintenance machine due to the acidity of the liquid manure. The farm has to buy a special slurry hauling wagon in order to transport the material—which is mostly water. Water is heavy. Transporting it compacts the soil in the fields during distribution.

The lagoon concrete cracks and breaks down due to the acidity of the manure. The slurry is so acidic, in fact, that it kills earthworms on contact. I've talked with farmers who spread slurry in the rain and the following morning saw the holocaust they created: millions of dead earthworms scalded to death, lying on top of the ground where they had tried to escape the burning.

A composting system is virtually trouble-free, does not utilize such specialized and high maintenance machinery, and can be done for pennies compared to a slurry system. And compost doesn't burn soil life. The differences between the two manure handling models were, and still are, compelling. The only problem was that mine had no official, credentialed recognition and the other one had all the authority of Virginia Tech's BMP writers behind it. That was a daunting reality, but I don't mind tilting at windmills once in awhile.

I decided to begin a one-man campaign for sensibility. I first contacted my local SCS guy—the one who funneled the money to the farmer and the one quoted in the newspaper article. He was supportive of my thoughts, but assured me that no money could come to us because we had already solved the problem. "You must understand," he said, "that grant money can only go to people who have problems. It can't reward people who have solved problems."

"So how do you decide what merits this incentive money?"

"We have to do research to show efficacy."

"Well, come on out and let's get started," I said. After all, we had a system up and running, proven over several years. What better place to analyze the effectiveness of a model?

"It's not quite like that. In order to study it, we have to study it in a controlled context. In other words, we can't research your farm because you're already doing it. We have to research it on a farm that's not doing any composting, impose a composting regimen on that farm, compare it to our current BMP, and get some baseline data for making a recommendation."

"Oh." What could I say to that? This is when science becomes its own worst stumbling block. It was similar to the time I heard the local extension agent extolling the virtues of antibiotics shot into the conjunctiva of the eye to control pinkeye in cattle. I called him and recommended our remedy: kelp. Dehydrated seaweed, rich in iodine. I told him we hadn't treated a case of pinkeye in years.

"I can't recommend anything that Virginia Tech hasn't signed off on," he said.

"Why can't you just say 'some people are reporting success with kelp' without endorsing it one way or another?" I queried.

"I'm not allowed to mention a single thing that hasn't been property scientifically studied and double-blind tested."

"Well, how do we get to that point?"

"Easy. Give me a big check to fund the research at the experiment station, and we'll do the study."

Indeed, we do have the best government money can buy. Again, dear reader, I am not making this up. These conversations happen thousands of times a day. The problem with the scientific method is that it cannot test multiple scenarios. It might discover the best among options A, B, and C, but if the real answer is E, and nobody ever thinks to test for E, truth never sees the light of day. The experiment can only be set up within the confines of the prejudices and paradigms of the researchers setting up the experiment. That is a significant limitation to the scientific method.

Back to lagoons. After I was deleted from research and deleted from incentive monies for an alternative to lagoons, I began a letter writing campaign. I wrote the governor and others, and finally scored a phone call from the SCS official in Richmond who was in charge of the entire farm lagoon program. His superiors decided to send him to Swoope to see this farmer who was creating such a stir. We set a date and he drove out to the farm.

He arrived on a cold winter day and I toured him through the deep bedded barn. He saw the content, warm cows lounging on their mound of straw and wood chips. It is always one of my most cherished sights, those cud-chewing cows relaxing on that soft, warm bedding. When we returned to the house, we sat in the living room and this is what he said:

"If you ever tell anyone I said this, I will deny it outright. But I will tell you that we have put far more manure into the Chesapeake Bay since this program began than we did before. What's happened is that now the farmers can store their manure for 6 months. In the summer they are busy with field work and other chores. They wait and spread it when they aren't busy, and guess when that is? In the winter, preferably on frozen ground, and often in the snow. Frozen ground doesn't let any of that manure penetrate anywhere. When the next rain or snowmelt occurs, every bit of it goes into the Bay. At least before, when they were spreading it daily, many of the days it was appropriate to spread it. But now they aren't spreading it on any appropriate day. It's all going into the Bay."

I was dumbfounded. I'll never forget that conversation—even though it's now more than 20 years old. I asked him why he couldn't use his clout as administrator of the program to change it.

Again, his answer was astounding, "You don't understand. The environmentalists won a huge victory with this program. How do you think they would look if they had to admit they'd caused more harm than good? And do you really think I can take on the entire scientific community at Virginia Tech? I'd be laughed out of town. And the politicians that passed the legislation—do you really think they will admit they made a mistake? They would look like fools to their constituents. Once a program starts, no matter whether it's working or not, the inertia to keep it going is stronger than anything else."

"Even truth," I thought to myself. Our meeting was entirely amicable. He could not have been more complimentary about what we were doing and what he saw. He was truly impressed. But I know he left with a heavy heart. He had seen the truth, but the truth did not set him free. It meant until his retirement in two years he would continue to promote a program he knew was doing more damage than before. But that was his job.

The Chesapeake Bay Foundation expected it. The politicians expected it. And Virginia Tech demanded it. And my idea was just pure lunacy. No seed money for research. No credibility. I was just a nutcase. That's the ugly other side of government grants.

They don't solve problems. They just perpetuate a perception that something is being done. That placates the peasants, but doesn't solve anything. Meanwhile, the duplicitous American pays more taxes and wonders why all these new problems keep sprouting up. Problem perpetuation, after all, is what keeps the economy moving.

About this time, the poultry industry was expanding and centralizing, to such an extent that the manure became a waste problem. Farmers were applying more to their land than the soil could metabolize, and big rains turned our Shenandoah Valley into a huge flush toilet, sending all these extra nutrients into waterways and killing fish. What to do?

The scientists were ready with a solution: feed the manure to cows. Today, in the spring of 2007, farmers are still feeding chicken manure to cows. More on that later. But the best research showed that it was completely benign for the animal.

The man who owned the small slaughterhouse that we use quit buying local beef because he said he became tired of walking in his chill room and smelling chicken manure. Beef isn't supposed to smell like that, after all. The scientists said it didn't affect the meat at all, but they were and are wrong. Still.

Today's hot environmental manure management solution is to make methane. Today millions of dollars are funneled into CAFOs to put plastic covers over their government-gifted lagoons and turn them into digesters to create methane. The methane powers all the equipment in the CAFO and suddenly an environment-friendly farm is born. The depreciation on this, however, is huge. If it can't economically fly without government grants, then it doesn't economically fly. Period. Ditto ethanol plants.

Our composting system does make economic sense, and environmental sense. My question to the methane promoters is this: "Why don't you get the cows off the concrete?"

But nobody is thinking about the debilitating affects of the concrete. Everyone is focused on that pile of slurry in that lagoon and the escalating energy bill. Using our farm's deep bedded system, half the energy costs to build the infrastructure could be eliminated right up front, which reduces the entire energy component in the system. With such little energy requirement, methane is unnecessary.

I am constantly amazed at the time and energy devoted to solving problems that should never exist. Stan Parsons, founder of Ranching for Profit schools, used to quip: "We've become extremely accurate at hitting the bull's eye of the wrong target." What an astute observation. We've figured out how to pour enough concrete to put a whole herd of cows on a slab big enough to be scraped by machine into a lagoon big enough to produce enough methane to power it all, but nobody has asked: "Why must the cow be on concrete?"

Such a fundamentally profound question never gets asked. The technicians are all frantically working on their part of the solution, but nobody asks if the solution is somewhere else. And when they're all done, they stand back and admire it and all the politicians, environmentalists, and industrialists pat each other on the back for such bi-partisanship. What a wonderful show of unity, truly a can-do attitude.

One of our apprentices recently attended his first Virginia Farm Show, an exhibition put on by the agriculture community of the latest greatest everything. He was fascinated by the water-bed pads for dairy cows. These are soft cushions for the cows to lounge on in their loafing stanchions to protect their joints from the deterioration that inevitably comes from a life spent on hard concrete.

Looking around the room, for the first time he could visualize the whole system. Over there is the plow that destroys the sod in order to plant corn. And here is the genetically-engineered corn at $100 per bushel, to put through the corn planter over there. And here is the sprayer for foliar and herbicides, called a Hi-Boy, for the low, low show price of $140,000 (a show-stopper $20,000 off regular retail). Ahhh, here's the 12-row harvester, another $180,000 machine, with its cousins: a couple of silage wagons, more tractors, and blower. Don't forget the silo. And the concrete. And the augers. And the fuel. Now add the water-bed stall mat, the manure scraper, the government-funded manure lagoon, a slurry spreader, pump, and more tractors. A veritable playground for machinery, fuel, technicians, and bankruptcy courts.

Why not just let the cow go out onto the soft sod in the first place, self-harvest the perennial forage, park the tractors, melt the plow, sell the Hi-Boy, forget the concrete, and give the water-bed mats to the homeless. Unbelievable. Then it could be a playground for cows, farmers, children, earthworms, and wildlife. Doesn't that sound fun?

The inventions are light-years away from the real solutions. You see, the government can never be creative, because by definition it must satisfy 51 percent of the population. And the majority is never on the cutting edge of innovation. Any study of innovation reveals a common thread: it's really lonely out there at the breakthrough point. The early discoverers do not have majority backing; they receive sneers and catcalls from the majority.

That is why any program deemed meritorious of government grants will inevitably be behind whatever the latest and greatest developments will be. Furthermore, government solutions must always be couched within the context of the existing paradigm. And today's agricultural paradigm is that technology is the solution to everything. If it doesn't dance with complex techno-glitzy glitter, it simply can't be a viable option.

Plainness, simpleness, don't register on the collective conscience. A culture that worships ever-thinner flat screen TV's and raunchier video games will not be excited about simple, humble agricultural solutions. Fortunately, the government has not yet deemed composting illegal, or simple solutions illegal, but I fear that the time will soon come when these ideas are indeed criminalized. More about that later.

Chapter 10

Conservation Easements

This topic is lengthy enough to require two chapters. I realize that this topic does not put me in the category of illegal per se, but it's related, because it shows how government officials and their programs operate. My personal experience with these is huge, and relaying them adds to the credibility of my notion that government solutions are really not solutions at all.

I can't believe how many times I meet folks who want to do something innovative and the first thing they say is, "I need to find grant money." From what I've seen, grant money keeps the innovator from being hungry enough to be truly innovative. The greatest innovation is at the point of starvation. Cushioning the creativity doesn't do anybody any good. And it locks the innovator into a protocol that usually becomes obsolete a third of the way into the innovation. Better to just rock along and make do rather than look for government help.

In addition, the official finding of a government research project is never the defining answer. The real answer eludes these officials because every waking minute is jaundiced by how to perpetuate research seed monies flowing in from outside the system. Research always starts with private seed money; and the seed money taints the findings.

Moving on from the lagoon era, the new buzz became water systems. Once the lagoon money ran its course, the environmentalists began pushing to get the cows out of the streams and ponds. Protecting riparian areas is indeed a noble goal. The exception is generally in brittle environments where periodic high herd intensity exposure helps soften creek banks and reduce erosion, just like we see today on the Serengetti when huge herds of wildebeests cross a river. The temporary hoof excavation leaves gentle grassy riverbanks a few weeks later.

But that is a different dynamic than the conventional cows lounging in the creek and farm pond. Just like the manure handling paradigm, though, cost-shared water systems have a capital-intensive bias. Again, I will describe what we use on our farm in order to contrast it with what government programs install.

I love ponds. Louis Bromfield, environmental farmer extraordinaire, loved ponds. In fact, he wrote in his books that if all the farmers whose land drained into the Mississippi would simply build ponds, it would eliminate all the huge Army Corps of Engineers projects being done to handle flooding. He argued that thousands of small farm ponds up in the headwaters were the only effective means to control the flooding. Once the water hits the river, it's too late to get control.

Over the years, we've built a dozen ponds and would like to build that many more as money allows. The beauty of ponds is that it creates landscape diversity, holds surface runoff to reduce flooding damage downstream, and grows aquatic plants and animals. And anyone can look at the pond and know how much water is there. Unlike a well, which sometimes goes dry all of a sudden, a pond's water can be measured to know exactly how much is left if we begin using more than is going in. Most of the ponds on our farm are winter runoff ponds, not spring fed ponds. That means they fill during the winter and we draw off them all summer until they fill again the following winter. In severe droughts, we've actually drained some of them. When that happens, we clean out the nutrient-rich muck in the bottom, haul it up on the hills in the manure spreader, and wait for fall rains. It's a wonderful fertilizer.

Taking advantage of our farm's elevation differences, we built a couple of ponds up in deep valleys and the water runs by gravity to the fields. By keeping the cows out of the ponds, the hydrologic plants maintain purity and the water is quite good. We have 5 miles of black polyethylene water line networked around the farm, with 99-cent valves every 200 feet. This gives us pressurized water everywhere. We just hook onto the valve with a length of garden hose when we need water in a certain location. For cattle, we use full-flow valves on 100-gallon Rubbermaid water troughs.

On rental farms, we haven't had the luxury of the elevation on our home farm, so we install high pressure pumps to send the water through the network. We trench the main lines in with a Ditch-Witch. We create frost-free valves by cutting pieces of 18-inch plastic culvert in 16-inch sections, digging a hole big enough to accept the piece of culvert, and covering the piece with wide wooden planking for a lid. We mill the wood on our bandsaw mill. This is a lot cheaper than frost-free hydrants, and not as easily broken over by cows rubbing their necks.

These systems are quite affordable. We put the pumps in simple T-111 boxes and wrap them in heat tape to protect them in the cold. One pump sucks out of a cistern filled from a well, another sucks from a creek, and another sucks from a cistern filled by gravity from a pond. In each of these cases, the systems utilize portable light weight cattle water troughs that we can move around on the back of a four-wheeler. The beauty of this is that we never have to place the trough, which is always a high impact zone, in the same place. This spreads the manure around and keeps one area from being torn up. This also builds in enough flexibility to locate cross fences in different areas to diversify paddock sizes and locations.

In government water development systems, however, this kind of flexibility is not offered. The hot item in this part of the country right now is the Conservation Reserve Enhancement Program (CREP) whereby farmers take riparian-sensitive areas out of production in exchange for cost-share monies to pay for the deleted acres and installation of a water system. I can't begin to keep up with all the government programs, but be assured all sorts of different cost-share programs exist to help farmers develop water systems that get the animals out of streams.

Realize, of course, that farmers like me who fenced out our riparian zones decades ago do not get any help. Only people who have abused their water resources get government help or new landowners. And that is one of the inherent discriminations in these programs. But that aside, I have watched with dismay as these programs build costly infrastructure. They never use ponds. They use wells or creeks.

A well is not only expensive to bore, but often does not deliver adequate water for a large herd of cattle. If the system sources from a creek, the frost-free installation requires that some sort of boring go down into the bedrock for a flood-free installation. All of these systems need wiring. On many of these farms where I've seen these systems installed, a pond could easily be built for a fraction the cost to let the water gravity feed. No power, no pumps. Our system at home gives 80 pounds-per-square inch static pressure on 5 miles of line without a pump; no electricity; no switches.

But the worst aspect of these government-granted installations is that they do not use portable troughs. They only use permanent installations—back to concrete, steel, and stone. Because these installations are permanent high impact spots, they require extensive stone out-aproning to keep the area from turning into a mudhole. Of course, if any cross fences are used in the grazing program, they must be put up in the same spot all the time, which again creates ruts through continuous high impact zones. And finally, these sites harbor pathogens because they do not enjoy host-free rest. The cows must come to that central location day after day after day.

On one of our rental properties, the government put in a $50,000 system that is useless because it doesn't deliver enough water to any one trough to actually water a herd. That means it assumes a continuous grazing program because in order for enough water to be there, the herd must use all three drinkers. Otherwise the recharge isn't fast enough. That means the animals have to spread out over the whole place, which is the most anti-environmental way to graze.

The bottom line, then, is that the environmentalists who claim a wonderful victory in getting the cows out of the stream are installing systems that preclude environmentally-friendly grazing management. That in fact insure that the pastures will continue to be thin and weedy, and that the soils will not uptake the water they should. I am constantly amazed at the millions of dollars going into these systems while the hillsides remain overgrazed and the soils remain non-absorbent. Talk about straining at gnats and swallowing camels.

This is all about doing poorly and feeling good about it. I asked a congressional staff member responsible for writing the enabling legislation that started the program why it couldn't use portable troughs and ponds. He said portable systems did not enhance estate value. In other words, only permanent installations with lots of concrete and steel would add value to the real estate. Ponds and simple, portable systems would not.

As with other government grants, the engineers designing the systems have no clue about environmentally-friendly mob grazing practices. They are in the water business. And water comes from wells. That's why in college they took Seismology 102. They aren't thinking about aquatic diversity. They are plumbers, not landscapers. Just another example of how a BMP is inherently skewed away from holistic thinking, because compartmentalized technicians designed the systems. And the researchers have never head about Permaculture, a movement founded by Tasmanian game biologist Bill Mollison. They don't read the alternative agriculture press.

It's not a conspiracy; but it is a fraternity of thought that does not recognize simple, humble, natural. It's a fraternity that worships technology, power, and domination. In the process, the systems are inherently energy-intensive, high maintenance, and non-adaptable. The research never considered better options.

Here's a good example. I attended a series of forage conferences one spring when the new high-tech snake oil was an herbicide that retarded senescence in grasses. When grass reaches senescence, its palatability and nutrient density drop. The cattle don't eat it as well, and what they do eat doesn't do them as much good. On our farm, we maintain palatability with management-intensive controlled grazing. By moving the animals every day or so the grass stays more palatable because we can maintain a vegetative state. This is similar to over mature vegetables that become woody and tough. We just don't like them.

The Virginia Tech animal scientist giving the presentation showed average daily gain of steers on pasture treated with this herbicide versus the control, which received no treatment. It was a conventionally grazed, non-rotated adjacent pasture in which the stocking rate was low in order to not run out of grass during the summer growth slump. This is the normal grazing program, and automatically creates a senescence situation in the late spring, resulting in the animals being forced to eat that over-mature forage during the summer. Of course performance drops.

The charts showed enough difference in average daily gain, that spraying the herbicide would pay for itself in increased per acre meat production. I sat through this presentation once and held my peace. A few weeks later, I attended another one-day extension-sponsored workshop and one of the sessions was this same professor doing the same presentation. The second time was too much to let go.

"Did you compare the herbicide-treated regimen with a controlled-grazing program?" I asked the Ph.D. pontificating this new-found technological fix to an age-old problem created by lackadaisical management.

"No, that was not part of the parameters of the research," he said. Of course not, because controlled grazing companies don't exist. But chemical companies do, and they dump piles of money on the scientific community to run "ours against nothing" tests to confirm their efficacy. If you're constantly testing something against nothing, lots of times the something will win, even if it isn't a very good something.

Our county decided a couple of years ago that something must be done about the proliferation of multi-flora rose. This is a pesky thornbush that was imported from England by the USDA in the 1950s to create living fences. The problem was that the U.S. did not have the natural pests that kept the plant in check. It quickly spread, and has become perhaps the most invasive scourge in the mid-Atlantic region.

Fortunately, it does not have an extensive root system. But it will come back indefinitely even if mowed off once a year. The root crown simply gets bigger and bigger under repeated mowings. It can climb up into trees, extending 20 ft. in the air. The thorns are practically alive and nasty, nasty, nasty. Birds spread the seeds from the little berries they produce.

The county found some money in a cost-share program and offered it to farmers for multi-flora rose eradication. To set the stage, on our farm, we don't have a problem in our fields because the sod is so thick due to our composting and nature-mimicking grazing that the plants can't get established. We have made some customized long-handled mattocks that we use to chop them out of fence rows. They are not really a problem in the woods because they need full sunlight to get established.

For us, the only place they are a problem are in field edges. We clean up some field edges every year in a multi-year rotation. Since we don't graze the woodlands, the understory and healthy tree canopies control the bush there. It's a problem, and we certainly don't like this plant, but it is a manageable nuisance.

When I heard the county would pay for eradication efforts, my ears perked up. I should have known better, though. The only thing the county would pay for was herbicide. Nothing else. They wouldn't pay for chopping. And they certainly wouldn't give a tax break if you figured out a way to keep the pesky plants at bay. No, that would never do. The only qualifying cost-shared technique was herbicide. The program automatically boxed out any alternatives and refused to recognize any other techniques.

We took an overgrown field and converted it to pasture. But the multi-flora rose seemed impossible to kill, and we had these little shoots coming up every year. Finally all of us went over and spent a couple of afternoons with mattocks and chopped out every single one of them. We've had no problem since. But that doesn't qualify for anything. Only the solution that pours money into the industrial sector is acceptable.

Sometimes the prejudices from the government agents are beyond obvious. A case in point. I attended a seminar on fertility taught by a Ph.D. agronomist from Virginia Tech. He started out the presentation with this statement: "The three largest components of the soil are carbon, oxygen, and hydrogen. But we're not going to talk about those. We're going to deal with nitrogen (N), phosphorous (P), and potassium (K)."

I couldn't help but think, "If you would focus on carbon, oxygen, and hydrogen, the other three would practically take care of themselves." Too simple, you say? Too picky, you say? Okay, let's take a look.

Carbon is what feeds the entire decomposition cycle in the soil. The decomposition cycle creates humus and the entire organic matter profile. Organic matter is what holds water: one pound holds four pounds of water. When the soil gets dry, I don't care how much NPK is there, nothing will grow. The limiting factor more times than not is not necessarily NPK, but moisture. And this includes the ability soak up water fast in a thunderstorm.

Decaying carbon releases carbon dioxide, which combines with water and forms carbonic acid. If I want to know what elements are in a rock—how much boron, calcium, molybdenum, etc.—I can treat that rock with numerous reagents. I can use sulfuric acid, hydrochloric acid. But if I really want the most efficacious one, I'll use carbonic acid. The reason our soils are depleted, for the most part, is not that the elements are gone. The elements are there, but not accessible because no carbonic acid is being created to break out the minerals that are there. Decomposition is the key to unlocking those minerals

Now oxygen. Oxygen gives the soil life breath. This is why tillage adds such zest to a soil by fluffing it up to increase oxygen. But if we have good porosity through earthworm and bacterial activity, oxygen penetrates the soil without mechanical tillage. Roots that live and then die create tunnels for oxygen and hydrogen transfer.

Six percent of a plant's compounds involve hydrogen, according to <u>An Acres</u> <u>U.S.A. Primer</u>. Since hydrogen comes from the water, this element becomes a limiting factor before many others.

In the lecture, then, the most important things—and the things that the assembled group of farmers needed to hear more than anything—were neglected in favor of the same old same old tired NPK material.

In the afternoon of the same seminar, an agricultural economist did a presentation on when is it right to buy calves or sell hay. In other words, assuming a farmer has hay, is it better to buy calves to eat the hay or just sell the hay outright. As a rule, 400-500 pound calves are cheaper in the fall than in the spring. A farmer doesn't really have to put any weight on them during the winter for them to be more valuable in the spring.

The forward margin of course is built into the market price because over-wintering is expensive since it requires stored forage. And stored forage needs to be mechanically harvested, which means machinery expense and labor. The presenter had numerous benchmarks for decision-making. Obviously if the value of the hay exceeded the forward margin of the calves and the labor to feed them, his position was that you should just sell the hay and not feed the hay to calves.

I kept waiting for him to address the value of the manure, but that never entered the equation. Why should I be surprised? On our farm, we have never sold hay because we view that as equivalent to selling soil. If we have extra hay, we always buy or board animals in order to get the goodies that come out the back end of the animal.

I've always thought cows are the first perpetual motion machine. After all, they eat 28 pounds of hay a day and give me 50 pounds of goodies out their back end. That's a pretty beneficial tradeoff, I'd say. To me, selling hay is almost immoral because it takes solar energy-produced biomass from one location and moves it somewhere else. The soil critters that expected to reap the benefits of all that vegetable matter are deprived their due reward. The payback for their work was supposed to be the decaying carbon they helped create. If we sell the hay, we make the soil network miss their dinner. And that's not very nice after they've worked so hard. We don't even move hay around on our rental farms. If we make hay on a piece of property, we do everything possible to feed it there and return the manure to that area.

But in the world of the agricultural economist, manure is totally worthless. Since most farmers put no value on it either, he considered his presentation a complete look at all the variables involved in the decision. But I know the NPK value alone of a cow's daily output is worth a quarter. Multiply that by 100 cows and that's $25 a day, times 100 days is $2,500. That's enough to seriously skew the analysis, but it never entered the mind of the researcher.

The point is that for all his charts and graphs, his snazzy powerpoint presentation and complicated computations, he completely overlooked the most salient value of the discussion: buying calves generates fertilizer; selling hay depletes the soil. But he never thought about it. In fact, when the question-answer session started, I asked—very diplomatically, I might add—about the value of the manure. He thought a moment, and then agreed that it would have been a valuable component, but it never crossed his mind.

I won't bore you with more stories like this because I think these are enough to prove the notion that government reports, and expert analysis, rarely give the whole picture. They come from a prejudice, a political slant, that requires thinking people to question them.

I hope that these personal stories will help all of us understand that we seldom if ever get the whole story from experts. If I could get every American who reads a government report to think first, "Now I wonder what these guys manipulated on this report, or what they neglected to put in it?" before agreeing with or to anything that comes down from on high, this will be a victorious day for righteousness indeed. One more time: I am not a conspiracy advocate. But I do believe that most experts come from the same school of thought, the same worldview, and therefore approach every problem from the same perspective. That gives consistent answers, and answers that for the most part are wrong.

Chapter 11

Restaurants

"The chairman of the board told me I couldn't buy from you anymore," said the chef, his voice almost breaking. The phone call culminated weeks of negotiations with the board members of the gated multi-use development.

The exclusive development included about 1,000 homes, a golf course, restaurant, theater, lake, and fitness center. Many of these multi-dimensional developments are springing up around the country as people look for integrated, safe living models that offer recreation, entertainment, housing, and beautiful scenery.

Many of these upscale residents are foodies, and their affiliated restaurants seek out artisanal local producers. A century ago, resorts offered this kind of multi-dimensional experience. Over the years, our farm has serviced these restaurant accounts with pastured meat and poultry, but too often they end with a call like the one quoted above.

In this case, again, we are not talking about something that is illegal in the strict sense of the word, but rather a cultural perception. A sort of paranoia about not being credentialed. A fear of freedom.

This is aggravating because what generally happens is that an outfit hires a fantastic chef to come in and prepare all this artisanal food. The chef understands the relationship between the quality on the dinner plate and the quality that comes in the back door of the kitchen. The chef immediately begins scavenging inputs from the nearby countryside and meets rave reviews from the patrons. Often some good local press accompanies the launch of the upscale restaurant.

These chefs, many of whom have been trained abroad where acquiring farm-produced and processed foods is more common, know that staying away from government-accredited fare is the secret to really good stuff. A facility big enough to fund the government's infrastructure overhead requirements generally needs raw materials from producers too big to be artisanal.

As just one example of how the inspection requirements ruin an otherwise wonderful product, consider grass-finished beef. When a steer is slaughtered, the muscle tissue releases an enzyme called calpain. This enzyme keeps the fibers from shrinking, or tightening, and instead makes them relax.

Activated by calcium and only viable in ambient room temperature, this enzyme works for only a couple of hours after the animal dies. But if the fibers get cold, it shuts down. One of the biggest problems in the grass-finished beef business is tough tissue, which many experts have blamed on insufficient intra-muscular fat, or marbling. This fat is easy to create with grain feeding; hence, fecal feedlots.

Yet hunters know that very lean venison and elk is tender, with virtually no intra-muscular fat. What's the difference? The difference is that wild game usually stays out at ambient temperature for hours before being chilled. By the time the hunter field guts the animal, drags or carries it to a vehicle, and then gets it to refrigeration, the meat has been out for hours, allowing calpain its maximum tenderizing function.

Under government inspection, however, the regulations require chill down in a certain period of time. In fact, the carcasses must be in the chill room being blasted by frigid air within one hour of slaughter. An animal that doesn't comply is automatically discarded.

Fat insulates. The intra-muscular chill down protocol established by the inspectors errs on the side of the slowest-cooling carcass. That would be the fattest one. Since many feedlots are real or virtual subsidiaries of grain companies, cattle have been viewed in recent history not as a means to grow wholesome beef, but as a means to discard mountains of government-subsidized corn. The more corn the animals ate, the better. Carcasses encased in fat were considered the best.

In fact, the grading system for prime, choice and so on are based on fat. Begun decades ago when tallow was one of the most valuable products of beef, prime simply designated the animal that gave the most fat. It had nothing to do with eating ability or health. The fatter the better, because tallow was more valuable than T-bones. These grades were delineated by government and industry leaders before rural electrification, back when tallow was more valuable because it was the primary ingredient for making candles.

Incidentally, the same is true for hogs. Their primary value was not tenderloin, but fat for lard. In the days before transfatty acids, everyone used lard for shortening and oils. That is why in those days hogs were generally not slaughtered until they were 400 pounds or bigger. Just like in people, physiological maturity facilitates fat deposition, like love handles. It's not fat, just a big candle lighting the way for the next generation.

But in nature, fat is much harder come by. Wild animals do carry fat, but they certainly don't carry the kind of fat that hangs on grain-fattened beef carcasses, which can often be as much as 200 pounds. This blanket of fat insulates the carcass, slowing the deep internal cool down.

When one of my grass-finished animals is shoved into the chill room next to one of these fat carcasses, the internal temperature will drop much quicker than the next door neighbor with a 200-pound coat of fat. Remember that the chill down standards are written for the slowest-cooling carcass, not the fastest, or even the middle.

As a result, in order to get internal temperature of the fattest animal where they want it as fast as they want it, the regulations inherently chill the leaner pasture-finished carcasses down much faster. Too fast, in fact. The faster cooling deactivates the calpain, which stops the tissue relaxation, which creates tough meat.

Many farmers have told me that when beef is killed on the farm, the meat is much more tender than it is when killed at a slaughterhouse. Part of that may be adrenaline activity due to handling and movement stress before slaughter, but much of it is due to the over-quick chill down required to meet broad, generic regulation protocols. In this case, the regulations absolutely destroy the quality of the artisanal product. At best they discriminate against environmentally and nutritionally superior pasture finished beef.

Chefs know this kind of stuff. One more story and then we'll move on. I was showing my eggs to a chef once and he wanted to buy them right away. To be perfectly transparent, I warned him that in the winter they wouldn't be as deeply orange as they are in the green grass season. I didn't want to be accused of a bait and switch deal.

He immediately cut in, "Oh, that's no problem. In chef's school in Switzerland we had recipes for March eggs, recipes for June eggs, and other recipes for October eggs in order to accentuate the nuances of that particular season's eggs."

I stood there with my mouth agape. In the U.S., an egg is an egg is an egg. Can you imagine McDonald's offering a different menu item to accentuate the seasonal nuances of eggs? April Egg McMuffin, October Egg McMuffin. Along with the food, an attractive point-of-sale info-bulletin on earth-tone paper would explain the differences: "Note the whitish spikes of albumen around the edges, indicating a thicker albumen as we move into the winter." What a hoot!

I can just see customers responding to this new connection with their environment: "Wow! Far out, man! McDonald's is really cool." Can you just see a hip-hop McDonald's foodie culture? "Hey, dude, it's May 15—time for the late spring Egg

McMuffin! Let's go check it out." Enough of this. Back to our artisanal restaurant.

The restaurant owners, whether they be investors, stockholders, or whatever, hire a chef, then, who understands that real food does not come through accredited channels. He goes outside the conventional system.

This works quite well until someone discovers that something in that kitchen does not have a USDA inspected shield on it. Then the proverbial egg hits the fan, no pun intended.

"What? You mean all the food in this kitchen is not USDA inspected?" The board member trembles in fear. The average American is absolutely paranoid about circumventing the system, especially when food is involved. People are removed from any real food-farm connection to such an extent that they swallow anything.

I have yet to see a chef break through this barrier. In every single case I'm aware of, the chef either leaves or gives in and starts getting things from the government-approved food supplier. Here's why. The board members or investors are themselves inextricably linked to the establishment wisdom in their lines of work.

Bankers, attorneys, engineers—their lives revolve around credentialed people with alphabet soup behind their names. I never cease to be amazed at how many of these people spend long conversations complaining about their respective oversight bureaucracies. The last time Teresa and I got a bank loan, half the discussion time involved the loan officer complaining about the bureaucracy.

He had a mountain of papers for us to sign, disclaiming this, waiving a right to that, making sure we were aware of something else. He said it added huge costs to his job, created needless paperwork, killed unnecessary trees. You name it, he just went off, apologizing for all the inconvenience and reciting a litany of indictments against the bureaucracy.

But this same man, when he goes out to eat tonight, will go apoplectic if a bureaucracy stamp is not on his food. Why? I don't know why. I really don't. But I hope this book helps to break through such schizophrenic reasoning.

The people who hold the chef's purse strings go to the same country club as the bureaucrats. It's a fraternity of ideas and a fraternity of system. They might not be in the same political party, but on this they agree: food without a government stamp on it might hurt you. Never mind that the food that's hurting people has the stamp on it. And even if someone were hurt by local fare, it would only be a few people. One supermarket hamburger is an amalgamation of material from as many as 1,000 animals. But that is deemed safe and biosecure.

A hamburger from our farm, by contrast, only contains meat from one or two animals. The sheer mathematical probability of cross contamination, therefore, is reduced astronomically in a smaller facility.

The frustrating aspect of this is that for the most part, the inspections that the shield represents have nothing to do with things that can hurt you. They deal with eye appeal and labeling issues as much as anything.

For example, when we began offering pork sausage, we were working with a chef who had a wonderful recipe she liked. For the uninitiated, pork sausage is simply ground pork with seasonings mixed into it. But since something is added to it, rather than just being meat, it comes under a whole set of different labeling requirements.

Naively, I figured we could just take her recipe, go purchase the herbs and spices, mix them together, take a bag into the slaughterhouse and have them put it in the ground pork. I figured the instruction sheet would be simple: Add 1 cup seasoning per 25 pounds ground pork.

Oh, how wrong I was. Such a procedure was grossly illegal. First, we had to get the seasoning recipe approved by the USDA. Then it had to be mixed in an inspected kitchen. To top it off, the mix had to be placed and kept in tamper-proof bags. And all of this had to be verifiably bullet-proof. In other words, we had to prove via a paper trail and paper explanation that enough oversight and protections existed in the whole process to never make an error. Zero tolerance. And the security had to be such that if anyone did tamper with the packaging, such tampering would be automatically noticed before the mix reached the ground pork.

It was just impossible. We'd have spent a month of Sundays and all our savings to comply. Alternatives? Find a pre-approved recipe. We went searching and found a company that made seasonings—it is the world leader. One small problem: the seasonings contained MSG. I asked the company: "Do any of your seasonings not contain MSG?"

They said I would need to check with their vendor who put them together. I contacted the vendor and after much hemming and hawing they found four types. But I couldn't just buy those and put them on my Polyface labeled product. Remember, these were deemed acceptable by the USDA: they were put together in approved facilities in tamper-proof baggies and all the recipes were approved. But even with all that, I couldn't use them until I personally received USDA approval for my labels. I couldn't just copy the ingredients onto my label. No, that would just make too much sense.

I had to go to a USDA office with the approved certificates from the seasoning company, show them to a bureaucrat, who then approved them onto my label. Now look, folks, in the final analysis if someone wants to tamper with something, it's pretty easy. At the end of the day, this slaughterhouse is crawling with people of all sorts of backgrounds, religions, languages, hygiene, and politics. The only thing policing all these workers is . . . nothing. Really.

I can't dress a hog for a restaurant without wrapping a million-dollar quintuple-permitted agricultural-zone prohibited facility around it. Here's this monolith of concrete and rebar and stainless steel to guarantee safe food. But when I go to that facility, pick up the raw pieces, put them in coolers, and deliver them to the restaurants, nobody knows if I'm honest. Nobody knows if I cleaned the coolers or sprinkled cow manure in there. Nobody knows if I'm out to hurt somebody.

See, the fact is when it's all said and done, the whole system depends on personal integrity. And personal integrity cannot be policed, legislated, or inspected. It just is or isn't. The things that people worry about simply aren't being checked. And if someone wants to taint something—the big fear that the USDA puts out in all its press releases to keep the populace fomented and paranoid and ready to accept additional bureaucracy and give up freedoms—anybody can do it. Easily. Especially in the largest facilities.

Do you know how many tractor trailers full of food crisscross this country on lonely rural interstates at 2 a.m.? Forget the domestic transport. How about the shipping containers that arrive at ports every day? The point is that the USDA does not check, and has no plans to check, and cannot check, the things it points its bureaucratic finger toward, exclaiming: "Bioterrorism threat! Bioterrorism threat!" The whole safety net is a smokescreen.

The bottom line is that the chef's bosses who told him he could no longer use our product were putting their faith in a bureaucracy rather than a local farmer. If I had to pick one to trust, I'd put my money on the farmer any day of the week. Yes, I'm sure some farmers aren't clean. But have you seen industrial slaughterhouses lately? Give me a break. This side of eternity, nothing perfect exists, period.

This discussion is not about how to achieve perfection. It's about how to minimize the possibility of food-borne illness either from pathogens or purposeful contamination. Every system will break down somewhere. The question is where is the propensity for the biggest breakdowns. I suggest that the individual farmer-entrepreneur whose reputation is at stake, and who can't afford to hire Philadelphia lawyers on retainer to act as a veil of protection between him and disgruntled customers, has more vested interest to act responsibly than global outfits. A little beast is easier to tame than a big beast.

After telling me he couldn't buy our chicken and eggs anymore, the chef asked the obvious rhetorical question, "What's their [the owners'] problem? Do they really think I'm out to kill my diners?"

And that is a good question. We have this cultural prejudice against business, as if good business can't exist any more. Unfortunately, many businesses are guilty of questionable ethics. But that is exactly why we must keep an extremely open access for competitors who practice better values and offer more transparency. The only real check on unbridled power and corruption at the top is unbridled alternatives rising from the bottom. I know to some folks that sounds like trite old laissez faire economics, and that school of thought is blamed for the powerful and corrupt businesses that exist today.

I wish I had a nickel for the number of times I've heard people say, "Look what free markets have gotten us—Enron, Martha Stewart, scandal after scandal." My only response is that we have not had a free market in much of anything for a very, very long time. We haven't had it in utilities, medical care, education, insurance, agriculture, or pound cake making. You can't spit anymore without permission from some bureaucrat. The government involvement always favors the big guys and hurts the little guys. Always, always, always. Whatever that involvement is, and no matter how sincere, the little players take it on the chin every time the government gets involved.

Whenever I do a talk about marketing, I spend my time telling people how to grow a better product and how to sell it. But during the question-answer segment, invariably, the whole focus moves to "is it legal?" The stories people tell me about the raids from bureaucrats will make your hair stand on end. And everywhere would-be clean food producers are hampered, stymied, and petrified of getting crossways of some labeling or food police. It is absolutely the most significant reason, especially in the livestock sector, why our movement has not displaced more industrial farming systems.

Unshackle the private sector, the thousands and thousands of farm and foodie entrepreneurs dreaming to access their neighborhoods with better food, and it would be equivalent to putting a governor on the speed of the industrial engine. In other words, as real competition hit the marketplace, the unimpeded growth—indeed, the aided and abetted growth—of the global corporations would suddenly slow. I believe this would create a restraint that naturally comes when peasants are free to assault dynasties.

This populism moves forward by sheer force from the previously disenfranchised rather than from top-down legislation orchestrated by the elite against some other elite. For example, the way to control the power of Big Oil is to free up the home brewing of alcohol. Get the government out of the liquor business and energy would democratize.

Let anyone butcher a beef in their backyard and sell the steaks to their neighbor, and the 4 companies that control 80 percent of the US beef market would suddenly see their oligopoly vanish. Ditto for poultry. Ditto for pork. The only reason the food system has become more concentrated is because competition has been summarily denied due to these burdensome government regulations.

Ever since our family made a hamburger mixing one-third pork sausage and two-thirds beef, we've never made another all-beef hamburger for our own personal consumption or to entertain guests. The sausage adds some fat to hold the lean beef together better, and the sausage spices add zest to the patty. The extra juice makes it cook better and keeps it more moist. It's a to-die-for burger.

But we can't make them for our customers without jumping through a pile of regulatory hoops. That's considered processed meat. Realize that in the same room, at the slaughterhouse, we can have a tub of sausage and a tub of ground beef. They are both safe to eat. But it's illegal to put the two together in an amalgamated product.

Or get this. At our custom butcher, we grind up tongue and heart and add it to the ground beef. That stretches the beef, adds a huge amount of nutrition to it via the organ meat, and puts the often-under-utilized organ meat to good use. But at the federal inspected facility, we can't do that to our ground beef, because the government has decreed that ground beef may not contain any organ meat. You can't put organ meat in ground beef and call it ground beef.

Doesn't matter that we and our customers want it that way. Doesn't matter that these are both safe to eat. Doesn't matter that we can do it at the custom facility. None of this has anything to do with food safety. It's strictly a capricious labeling issue plaguing us from the dark ages.

I was speaking at a conference in Alabama, and a fellow leading a statewide effort to get local food into school cafeterias told me that the internal policy of the lunch police dictated that "you can't put a fresh piece of vegetable on a plate of school lunch without putting it on every plate in the state for that day." He said that total was 463,000 plates. In other words, his efforts to access one local school cafeteria were stymied and stonewalled because he couldn't get the item to every single plate in the state on the same day. That's ridiculous.

It reminds me of when I used to do a Bible study at the local prison and I found out that the dining hall could not feed leftovers. If some prisoners received fresh chocolate cake for dessert while others received yesterday's chocolate cake, that was deemed unfair. As a result, the kitchen summarily discarded leftovers in order not to sleight anyone. Give me a break.

Here's another good one. We can sell our on-farm processed uninspected chickens to restaurants, which of course can store them in a freezer or cooler. But if the restaurant buys any poultry from any inspected source, those chickens cannot be stored in the same refrigerator or freezer. I believe this is to keep our clean birds from being contaminated by the inspected birds.

What this means is that if we are too small a producer to supply a restaurant, but a chef wants to get as many of ours chickens as possible, conventionally-sourced chicken cannot be stored in the same units as ours. This obviously puts the restaurant in a huge bind, and it's just easier to not run afoul the chicken police by getting the industrial inspected birds—from 1,000 miles away. By what stroke of lunacy did this policy develop? If the birds are safe to eat, who cares if they are co-mingled in the freezer?

Are the bureaucrats afraid that a local farm bird and a distant factory bird might get together down there in the cold, dark freezer and perform some hanky-panky? Either the chicken is safe or it isn't. If one is unsafe enough to taint the other, then the unsafe one should be prohibited from the premises. The average

American cannot imagine the regulatory minutiae accompanying every food transaction in this country.

The cumulative effect of all these requirements is that the local producer stays nonviable as a business and as a player on the world food stage. Were local access encouraged rather than discouraged, the world food stage would at least be significantly altered by the thousands upon thousands of local venues. As it is, the local food system doesn't even register in Washington D.C., or even at the state level. It's a virtual nonentity, and the brunt of jokes.

Local food advocates like me receive jeers from the credentialed agriculture community as elitists, as irrelevant, as a joke. I'd like these regulations to be lifted for just one year. At the end of that year, I guarantee these jeering, leering corporate and university fat cats would be reeling from the power of an unleashed local food system. The fact that a local food system exists at all bespeaks its potential power. If we can do what we've done with both hands cuffed, imagine what we could do were we freed to play unimpeded. We should all dream about such a day. Let the revolution come quickly.

Chapter 12

Predators and Endangered Species

I like wildlife. Even though I'm surrounded by chickens, turkeys, and cows, my heart still thrills at a deer prancing across the field. I think it's because no matter how many domestic animals we produce, we're still in charge of them. The wild ones can't be controlled. We can't get them in a corral whenever we want to. If we have a hankering to see a Red Cockaded Woodpecker, we can't just walk out to the field shelter and look at one. We see them on their terms, not ours.

With that said, I experience an equal thrill when finally exterminating a predator that has picked my pocket for a few nights. Unfortunately, predators are considered wild animals, and wild animals are state property. Some of the most rapacious predators are still protected as endangered species.

Perhaps the place to start on this topic is with an airplane encounter with a leading wildlife advocate, a member of the environmental movement's elite inner circle. I was flying back from a seminar and we began chatting. He had recently been to the White House Rose Garden for the hysterica . . oops, I mean historical signing of some environmental-friendly legislation.

Predators and Endangered Species

At the risk of destroying my credibility as an environmentally friendly farmer, let me just make a point before I go on with the airplane story. According to folk tradition, somebody asked multi-millionaire J. Paul Getty one time, "How much money is enough?"

To which he replied, "One more dollar."

I think this is an unspoken rule more often than not, in a lot of different areas. For example, I'd like to ask Bo Pilgrim, of poultry giant Pilgrim's Pride, "How many chicken houses are enough?" Everything in their business is geared toward expansion. Here in the Shenandoah Valley we have Cargill, Pilgrim's Pride, Purdue, and Tyson, and all of them want to build more chicken houses. Expansion represents success.

I can just hear all the liberals and environmentalists yelling "Yeah! Preach it brother!" Okay, let me ask you something, "How many wilderness areas are enough?" It's the same mentality. It's expansion for the sake of expansion without understanding that for every action there is an equal and opposite reaction. And lest you think the roots of this expansion began with the Biblical admonition to populate and dominate the world, I would suggest that Jesus' mandate to teach all nations and make disciples carries an override protection. Making disciples takes time. Disciples come slowly.

When the Shenandoah National Park displaced hundreds of Appalachian farm families, sending them to the lowlands and into cities, the social cost of that forced migration was just as devastating as the Trail of Tears to the Cherokees. Those who would castigate what we did to the Indians—and I am one—seem to turn a blind eye to the whole communities that our society exterminated to make way for national parks and wilderness areas. Dispossessing whole communities is a disgrace. These areas, rather than supporting a heritage culture and honoring the mountain wisdom of these subsistence farmers, produce nothing but trees. Trees that can't even be harvested. Trees that grow up, get old, sick, and die. Decaying trees that become net producers of carbon dioxide rather than vibrant, growing trees that create oxygen out of carbon dioxide.

Out West the fuel build-up in the no-cut areas is a resource disgrace. Before modern fire suppression capabilities, wildfires acted as nature's chain saw to control fuel buildup in the forests and to kill old trees and rejuvenate the landscape. Indians routinely lit wild fires to beat back the forest and create open savannahs to feed and attract bison. In the blockbuster book <u>1491</u>, author Charles Mann meticulously documents the early Americas as heavily populated, highly managed, and thriving with commerce. With modern fire suppression, we're growing more trees than ever.

In 1820 Vermont was 20 percent forested and 80 percent open; now it's exactly the opposite. Our county had 50 percent fewer trees in 1860 than it does today. I've been to the redwoods and they are awesome. But I get just as excited about seeing the mid-growth virulent trees as the old matriarchs in the national parks. A tremendous amount of latitude exists between a clear cut everything policy and a don't cut anything policy. Both the tree huggers and the foresters have points to endorse and points to throw out.

The environmentalists have practically criminalized clear cuts in this area. I love little two-acre openings. They bring on blueberries and lots of diversity. But that's a far cry from a 100-acre clear cut. The devil is in the magnitude. Lots of things, when performed with restraint, become big problems when done to excess. Drinking comes to mind. And by the same token, just because a two-acre clear cut is great doesn't mean a 100-acre one is better. Both sides need to stay in balance. I welcome anyone who doesn't believe me to come and see what we've done over nearly half a century with our woodland. It's a living, growing organism that needs care, just like when the Indians started fires that burned thousands of acres. Except we use chainsaws and lumber mills.

Until recent years, when the forest service wanted to harvest some timber, the area was marked and offered as a clear cut. The logger moved in, took down everything, and moved off.

The size of the areas was arguably too large. But today, because clear cutting has been demonized, these same cuts on government land are offered as thinnings.

In a clear cut, every stem larger than an inch or two in diameter is harvested. A tree too small to be marketable must still be cut in order to have a clean tabletop. The clean tabletop is what simulates fire and creates an equal opportunity regrowth period. It's also the key to a full sunlight penetration to the floor that creates the full successional cycle, from brambles to pioneer species to the longer-growing trees. All natural systems have a succession pattern of early period and mature period growth. Forests do too.

The forest you see today will not be there tomorrow. The skin you see today on your arms will not be there tomorrow. The nursing home occupants you see today will not be there tomorrow. It's the nature of living things, and forests are living things. A no-cut policy in the forest is tantamount to a no-bathe policy for people. And the book 1491 now proves in graphic scientific detail that so-called wilderness areas never existed. Be assured that the wilderness areas that exist today with official designation will not look the same in a century. You simply can't capture a living thing in a moment of time and expect all cyclical maturity to stop.

The Shenandoah National Park is the first to suffer from air pollutants, gypsy moth infestations, and other tree problems because of its no-cut policy. The dead, down, and dying timber in those thousands and thousands of acres are a disgrace to our culture, and an immoral waste of solar energy and God-given renewable resources.

In a thinning, the logger takes only a portion of the trees and many are left standing. Certainly thinning is a viable silvicultural technique, but only if done judiciously and meticulously. And only certain stands are conducive to this technique. As currently practiced on government lands, at least around here, it is a disaster. Too many trees are cut, and not enough good trees are left in the understory.

Within two years, the remaining trees bend over, become hairy due to epicormick sprouting, get wind shook, diseased. The tops fall out when the wind comes through. And many die due to scarring that occurred during the initial harvest. As a result, two years later the same site is marked for what is called a salvage cut.

Now rather than high quality trees, everything is junk and goes for bottom prices. Loggers don't bid nearly as much, of course, because the residual material is spread out and makes harvest inefficient. The point is that rather than having a one-time move-in efficient harvest, it's a de facto clear cut that took two years to accomplish. This happens all the time, and it's crazy. Why can't they strike a balance between the two extremes? Harvest in clear cuts, but in tracts 10 acres or less. Then you have the best of both worlds with none of the negatives.

The reason that can't be done is because the two camps are polarized. I know, because I have wonderful friends on both sides. The professional foresters know, and I say that guardedly, that 100 acre clear cuts are fine. And they can recite all sorts of silvicultural information to show why. And they are friends of the loggers, and the loggers want it big. Bigger is better, remember?

The environmentalists know, and I say that guardedly, that clear cuts are anathema. Period. End of discussion. Zero tolerance. And ne'er the twain shall meet. That's real life on the land, USA, folks. It happens every day. And neither side will concede an inch. The environmentalists can't concede even a two-acre clear cut because in their lexicon anything that disruptive is sinful. How about a townhouse? How about a house and lot graded into a cul-de-sac in a former woods? How many tree huggers will give up their homes? How easily we become sanctimonious—with someone else's decisions.

Now back to my environmentalist airplane conversation. Our discussion finally turned to food, and I promoted the idea of "getting in touch with your food supply."

He said, "I don't have time to get in touch with my food supply. I live on Nabs and vending machines. I'm too busy saving nature."

After we landed and went our separate ways, I thought about that statement. What a perfect example of the disconnect in so many minds, especially the policy wonks. The world certainly doesn't lack for people who try to use the power of government to make other people do what they think they should do.

I remember growing up during the DDT era, when we never saw a hawk. They simply didn't exist. Then in the late 1970s we began seeing them. Today, one is perched on every fencepost lining the interstate. They are everywhere. Up until just a couple of years ago, Canada Geese were listed as an endangered species on our side of the Blue Ridge Mountains and on the east side the state paid a $10 per head bounty to kill them.

As my son Daniel would say, "What's up with that?" Yet the same people who would haul me to jail if they saw me shoot a hawk (notice, I didn't say I ever have—I'm just postulating this to move the discussion forward) think that all chickens should be pastured like ours are. Think about how much you like pastured chicken. Savor the thought. Now multiply it about tenfold and that's how much a red-tailed hawk likes them.

The pastured eggs our farm produces are the best on the planet. No question. One of the reasons they are expensive is because of the time involved in providing security. Here's my question to the environmentalists, "How many hawks are enough?" Folks love to use this question against profiteering corporations—like "How much market share is enough?" or "How much profit is enough?" But they don't have any problem demanding one more hawk long after the cultural shift against DDT created a wonderful and exponential comeback. Without any Indians to trap and kill them, the hawk predators are gone. I think we need to bring back the hawk feathered headdress. That would create commercial hawk value and people would begin hunting them again.

You can't take humans out of the landscape. We are part of it, like it or not, and have been for a long, long time. This notion that humans are inherently damaging to the landscape is simply an over reaction to the damage inflicted by humans. When I promote getting more loving stewards out here on our farm to live on it and capture more solar energy through additional vegetation, greenies give me that quizzical look that says: "But more people will destroy it."

No, they won't. Not if they adhere to natural principles. We don't come close to producing what this land could produce, even while it heals. One of my icons, Wendell Berry, makes the excellent point in his classic The Unsettling of America that ultimately the rabid environmentalist and the rabid factory farmer are cut from the same cloth: they both idolize a landscape devoid of humans. Ultimately, they both hate people. That is a shame, and should give us all pause. Asked to supply a picture of the ideal landscape, neither group will include humans in the portrait.

This whole issue is driving the Buffalo Commons idea in the west, as if the cow is the problem. The cow is no more a problem than the buffalo. It's all about the management. Don't blame the cow for a people problem. Ranchers who understand these natural grass growth and recovery principles, espoused perhaps most articulately by Allan Savory and Holistic Management headquartered in Albuquerque, New Mexico, spin circles around the federal bureaucrats in their understanding of nature.

But the Bureau of Land Management policies routinely do not allow the kind of innovative cattle management practices that heal the land. Environmentalist-turned-friend-of-the-rancher Dan Dagget is probably as good a spokesman for this conversion experience as anyone. His story is both compelling and entertaining. Nobody has a greater distaste for what poorly managed herbivores have done to the West than I do. But the desertification of the West is not an animal problem; it's a people problem. Herbivores can harm or heal; it's all about the management.

Removing ranchers will not solve a single problem. Human activity is part of the natural landscape. Human activity is not inherently evil toward nature. If ranchers bear substantial responsibility for land deterioration in the west, certainly the federal BLM agents bear an equal share. Thrusting BLM agents into the landscape is no more protective than thrusting ranchers in. All the BLM agents do is create roadblocks to the good ranchers who do want to innovate and heal the land. All ranchers are lumped together as bad.

Predators and Endangered Species

Rather than getting out of the way to let these innovations occur and proliferate in a normal informed leader-follower sequence, the federal presence obfuscates, obliterates, and adulterates. Here on our farm the federal presence in the Shenandoah National Park and the state presence in the Virginia Division of Game and Inland Fisheries has created a bear problem. If neither of those agencies owned land, the native bear populations would go about their business. But the way it is, all the campsites in the park are attractive nuisances to bears. Park officials routinely trap bears and relocate them to the game commission properties, which border our farm.

The bears then traverse the Valley's farmland to get back home to the Blue Ridge Mountains on the other side of the Valley. This constant government manipulation of the bears creates problems for us farmers who have to deal with these extra-clever, people-smart bears. Especially farmers like us who have chickens in the pasture or pigs on pasture. While environmentalists locally have asked for more and more wilderness areas to reduce alleged "fragmentation" of the forest, these wild animals become more adapted and more clever. This creates more havoc for farmers, who find it harder to compete and stay in business. The resultant abandoned farmland gradually reverts to woods, creating more habitat for the bears and putting more pressure on adjacent farmland.

I am certainly not a bear exterminator. I enjoy seeing a bear from time to time. But I've also seen our pastured pigs after a bear attack. A 150 pound pig dragging around its hind legs because a bear swiped down and broke its back is not a pretty sight. A century ago, the whole community ran its pigs in a big group on the mountain. In the fall, the group would be enticed to a corral with corn and each farmer would take out his own pigs—identified with little ear notches.

We would love to do that today, but allowing a pig over onto a neighbor's land or the government land would be inappropriate—and illegal. Hence, the acres and acres of acorns that could fatten the hogs go uneaten while we buy corn as a substitute. Meanwhile the protected bears continue to escalate in numbers, cleverness, and boldness. I was up one morning early to

165

move the Eggmobile and there was a bear trying to rip the steel roofing off to get in to the chickens. I chased him off, but I really wish I'd had the 12 gauge loaded with buckshot. That would have put a little natural balance in our ecosystem—at least until the next park bear got dumped next to our farm to start the escapade over.

To many Americans, farmland is a big park. It's Recreational America rather than rural America. Many folks in urban areas have no clue that people actually live here and battle blizzards and hawks and bears and drought. We actually go out at 2 a.m. to help deliver a calf or save chicks from rising water. And farmers like me, who base their livelihood on pasture and not CAFOs, are more susceptible to these invasive ideas and policies. We're trying to co-exist with nature. But when other farmers pen their animals up in a fecal factory house, we pasture-based outfits are the ones most vulnerable to protected and manipulated predators.

One year we had a particularly wily fox digging under our field shelters and eating broilers. Fortunately the chickens were within rifle shot of the house. I began getting up at odd hours of the night, shining the spotlight out there to see him. I only saw him once. If I was early, he'd come late. If I was late, say 3 a.m., he had already been there. I finally spent a couple of nights in a tipped-over feed tank out by the birds. He still outfoxed me. I could hear him yapping at me over in the woods.

I finally get a couple of shots off at him one night—actually thought I got him—but a couple of nights later he was back. Then it struck me: the state owns all the wildlife. If the state owns all the wildlife, they should be liable for a miscreant doing me damage. I thought this was a pretty clever tactic. I called the Virginia Division of Game and Inland Fisheries: "One of your foxes is in my chickens. Could you kindly come out and control him?"

"Well, sir, you misunderstand the relationship here. This is a partnership. The wildlife is half ours and half yours."

"Ah, then whenever I get low on venison, I'll just go out and shoot my deer. But I won't shoot yours; they can stay until hunting season comes in and your people can take care of those."

"Well, it doesn't really work that way either. Ownership might be too strong a word. We have guardianship over all the wildlife, like custodial responsibility."

"Okay, then when one of your deer runs out and destroys the front of my car, I'll send you a bill. Kind of like parental responsibility for a minor."

"Well, individual landowners can't own the wildlife, but the state is still not responsible. I think I like the half and half ownership idea."

"I'll tell you what. I've worked real hard to control my half of that fox and I think I've been successful, but his other half—the half you own—drags my half around all the time and causes trouble. Now when hawks come, I only deal with my half of the population. This fox is becoming intolerable, and I'd greatly appreciate you getting your half under control."

This was all done tongue-in-cheek and he enjoyed the banter as much as I, but for any thinking person, it certainly shows the inconsistency and logistical impossibility of the state owning all the wildlife all the time. The game commission sent out a trapper and he couldn't get control of his half either. But finally the fox felt enough pressure that he moved on to other areas.

This summer we had a terrible predator in the chickens. He was getting in the shelters and killing half a dozen a night, big ones. It's not as bad when the birds are small, but when they are big it's a costly loss. He didn't eat much—maybe a head or two—but seemed to just enjoy killing. Daniel and the apprentices took turns getting up for several nights but couldn't get him. We set a land mine of leg traps. Finally the old man—me—decided to solve the problem.

The birds were near one of our hay sheds, on the other side of a brushy fence line. I took the chainsaw and cleaned the fence line and then announced I would spend the night in the barn. Due to the magnitude of the slaughter each night, I assumed that I would hear him when he arrived. I took my .22 magnum bolt action along with a pump 12 gauge shotgun with buckshot. With the two guns and spotlight, I was ready for anything.

I crawled up in the hay and actually went to sleep. About 10 p.m., a boisterous banging and ruckus suddenly signaled the arrival of our unwanted visitor. I got myself positioned with the two weapons and flicked on the spotlight. Nothing. The chickens were going crazy in the shelter, flapping up against the metal sides and squawking up a storm. I still couldn't see anything.

I said to myself, "Okay, you devil, if you won't come to me, I'll go to you." I climbed down out of the hay and sprinted the 50 yards over to the shelter where the killing was obviously going on. I ran around to the front and there he was, inside, killing chickens as fast as he could. Two were already dead on the ground. When I hit that shelter with the spotlight, all the metal siding reflected and lit up the scene like the stage in a theater. I raised the shotgun from about 15 feet and put a charge of double-aught right into his neck.

Oh, did you want to know what it was? Do I dare tell? It was a marten, close relative of the ferret and weasel. I enjoyed the distinction of mighty hunter when the boys got up the next morning.

I could regale you with lots of stories after 40 years of pasturing poultry: that's a lot of succulent dinners for our patrons over the years. And we've stood guard, trapped, shot, and brought those dinners to the pot. I don't know where all the answers are in finding the balance on these predatory animals, especially the ones that are still on the endangered species list.

Some people say we should use a dog. Okay, but we have to train it. It has to stay with the chickens, especially at night. It can't be lounging around the back stoop when the great horned owls are descending on the chickens. And we have such a number of customers, often with children, that we wouldn't want a vicious dog. A dog vicious enough to chase off a coyote might not be the most benign around strangers.

We have found geese to be fairly effective at chasing off aerial predators in the Feathernet. The Feathernet is an oval of electrified fencing around a portable hoop house shelter. The laying hens free range inside the protected oval, but are vulnerable to aerial attack. But when we try to put the guard geese with the totally unrestricted birds at the Eggmobile, we can't make them stay there. They always wander down to the closest pond and don't come back. Ditto for when we put them with the broilers.

Predators and Endangered Species

Where I'm going with this is that if we didn't have to keep looking over our shoulders when dealing with the few rogues, it would not be nearly the problem that it is. The gross illegality of taking out a hawk is what makes the whole issue difficult. We certainly don't have an attack every day, or even once a week. Fortunately, with all the human activity around, our presence creates a sort of buffer that most farms don't have.

People love to come and take pictures of the pretty chickens in the field. Journalists love to come and sing our praises, describing in beautiful prose the magnificence of pastured poultry. It would be much easier to do this if what we really needed to do weren't illegal.

And one final note about a less wild predator: the domestic dog. Although we live way out in the boondocks, many pastured poultry producers do not. In fact, we promote our agriculture as being aesthetically and aromatically pleasant enough to be imbedded into a community. Too often this community has a few Fi-Fis that turn into marauding rogues when let out to go potty at night.

As more housing developments crowd farmland, pasture-based agriculture again is most vulnerable to every negative human extension. And dogs are certainly a big one. Our own local newspaper has carried numerous stories of some poor farmer drug through the courts, sued by a dog owner who lost a pet. Of course, the pet was trespassing on the farmer's land. I've seen a flock of sheep dragging their intestines, bleating pitifully, after a couple of pet dogs indulged their instinctual desires.

The problem here is that for the average American, far removed from food production and farming, the only human-animal connection is pet oriented. The predator-prey relationship is viscerally existent out here in the hinterlands. And the average person has no attachment to domestic livestock to temper an almost idolatrous worship of the pet. For us farmers, a pet is just one of many animals on the farm. It's not the only critter to ogle over and enjoy our affection.

Now for the confession: whenever a dog comes on our premises, unless we know it to be a neighbor's dog, we practice the rural three Ss—Shoot, Shovel, and Shut up. Every farmer in this area practices that protocol, and it maintains a balance. If that sounds harsh, so be it. But by the time you call the Dog Warden, the marauding critters have high-tailed it up over the far hill and are long out of sight. The time to deal with trespassing marauders is at the point of impact.

As dog protection laws become more imbalanced, I see this as a larger looming threat. Dog owners have the responsibility to control their animals, just like I must keep my chickens and cows at home. The very thought that it may be illegal to shoot a dog running through the field with a chicken in its mouth makes my blood boil. But I'm afraid those days are coming with the increasing self-absorption and food disconnect suffered by mainline Americans.

Again, the greatest inconsistency of this is the people most vocal about protecting animal rights are the ones who most appreciate pastured poultry, pastured pork, and grass-based beef. In other words, the folks driving the pet protection agenda and the ones most quick to sue a farmer like me tend to be our most ardent supporters. This is just another glaring example of how most of us become myopic in our thinking, our action, and our policy agenda.

The conservative groups most vocal for property rights, who would fight for my right to kill a marauding pet, tend to be the least likely to care about eating environmentally friendly and more nutritious food. It's one of the great anomalies of our day, but it is true. It's actually maddening sometimes. I hope this book will help create some awareness and connection where it was sadly lacking before.

This is a big problem everywhere in America—wherever people put poultry out on pasture. Unfortunately, the average foodie has not made the connection between the successful pastured egg and the controlled hawk. Again, everything I want to do is illegal.

Chapter 13

Sawmills Are Out

The public hearing announcement for zoning ordinance amendments caught my attention in the classified ad section of the newspaper. In journalistic jargon, the heavily gray ad—meaning lots of writing and not much white space—enumerated half a dozen proposed amendments to the county's zoning ordinance.

Although I don't routinely feel compelled to study these announcements, the word "sawmill" caught my attention in this one and I stopped scanning long enough to read it. The proposed ordinance change outlawed sawmills in agriculturally zoned areas. Our farm, like most in the county, lies in an area zoned agricultural. The point of the ordinance is to prohibit land uses that society believes conflicts with farming.

At the time, we were hauling logs from our woods to a neighbor's bandsaw mill about five miles away. We'd haul them there on a hay wagon, and half the time we'd have a breakdown on the way over. Of course, we always loaded the wagon too heavy, trying to get the most out of the trip. Hay wagons are called hay wagons because they are designed to be loaded with hay, which is much lighter than logs, and pulled behind a tractor, slowly, across a hay field.

They are not built to go 30 miles per hour behind a pickup on the public road loaded with 4 tons of logs. Invariably we'd suffer a flat tire, broken hitch or some other calamity on the way over to the neighbor's mill. But it was better than nothing, so that's what we did. And even with all the problems, it was still much cheaper than buying the lumber at a commercial mill.

But we desperately wanted a mill of our own. Bandsaws had been out long enough to prove their viability and enough entrepreneurs were entering the mill manufacturing business to bring the prices down. We had our eyes on a couple of models, and were eagerly accumulating enough capital to buy one. And now suddenly I was looking at a proposed ordinance that would outlaw my dream machine.

I made some phone calls and determined that the reason for the proposal was that a couple of commercial sawmills in the county were being terrible neighbors. Hiring lots of illegal aliens and cramming them into glorified chicken houses, heavy trucks entering and exiting at all hours of day and night, noise pollution, light pollution, and clouds of dust wafting across the neighborhood—everyone knew that a sawmill was commercial and industrial, not agricultural.

I don't know how the ghastly fact that sawmills were legal in agricultural zones formally came to light in the consciousness of the county zoning bureaucracy, but someone there discovered the egregious freedom. The road that leads to these discoveries is often mysterious and never straight. It usually starts with a complaint from some well-meaning citizen who wants the government to do something. Possibly, some farmer with a mill got on the outs with a neighbor, who then complained. Many times these things start because two neighbors can't get along.

However this gross oversight was discovered is not as important as the fact that in the wisdom of local zoning administrators, the proper solution was to prohibit sawmills on farmland. The amendment, like so many of these, mentioned nothing about milling for personal use, or size of the mill, or number of employees for the mill. It was very simple: all sawmills. Regardless of scale or use. Period. End of discussion.

To help set the context for what faced us, personally, our farm has 450 acres of Appalachian hardwoods. In many ways, although our farm is famous for pastured livestock, it is really a forest farm. And we knew the forest offered many income opportunities, if we could value add it by converting logs into lumber. I had been cutting firewood for a long time, upgrading little areas by culling out the diseased, crooked and dead trees.

Eventually, though, the good trees reach maturity and must be harvested before they get diseased and fall over. With the building projects we had in the offing and the forest quality steadily improving, we as a family entertained serious dreams about the profitability of sustainable forestry. Teresa and I thought about using a mill to build inexpensive drying sheds, sticker walnut and cherry and quarter-sawn red and white oak, and letting that air-dried lumber be our Social Security program. We reasoned that in 50 years when the Nature Conservancy finally silenced the last chainsaw, perhaps our little stash of boards would be pretty valuable.

Lumber is an absolute staple on a farm. Building and maintaining outbuildings requires lots of lumber. And buying it from Lowe's or Home Depot is simply unaffordable for many projects. As a result, most farmers don't even start on innovative projects because the lumber cost prohibits the dream from ever becoming a viable alternative. Corrals and sheds are just too expensive to build and maintain when all the wood comes from the hardware store. As a result, most farmers continue to patch their dilapidated corrals and outbuildings together with baler twine and wire.

In our multi-year quest for affordable lumber, we actually tore down a couple of barns in the community to salvage the boards. Even though these buildings had deteriorated beyond the point of repair, they still had plenty of usable lumber. We re-used the sound material and burned the junky stuff. In fact, I used some beautiful weathered exterior sheathing boards to build Teresa and me our clothes closet in our bedroom. It's my crowning carpentry project. Our old farmhouse doesn't have any built-in closets. To add any would be illegal without a license. But a free-standing closet is considered furniture. That's actually legal. Appreciating wood, using it all the time, and trying to find cheap sources consumed our imagination.

When portable bandsaw mills came out, I would covet them, not in a greedy way but in the way that someone passionately seeks a solution to a problem. We had a problem. On the one hand, we had a steady and necessary appetite for lumber for our numerous farm projects. And on the other hand, we had nearly a square mile of prime upland hardwood forest staring at us every day. To sell it wholesale was simply out of the question.

I had already tried that, and vowed to never do it again. One winter I cut good logs and towed them with the tractor to a flat spot. When I finally accumulated a truck load, I called a logger who had a knuckleboom loader to haul them to a nearby commercial sawmill. He came over, loaded the logs, and took them to the mill. I rode with him. When we arrived, he unloaded all the logs side by side, not stacked up. Then a fellow came out of a little building—called a scale house—carrying a log scale. He measured each log and computed the number of board feet (12 inches X 12 inches X 1 inch) contained in each one. He graded the logs as to A, B, or utility. Obviously, premium grade logs fetch more per board foot than utility.

I had some nice logs. Clear (without knots) red oak, white oak, some beautiful poplar. They were straight, very little taper (the difference between the big end and little end), and sound (no rot inside). The problem? Since they had been cut over a period of two months and weren't in pristine, green condition, most of them went to utility grade. Even though the wood was just as good as if the log had been brought in fresh-cut and green, that little weathering and age dropped them three grade levels. A log that would have brought $150 brought less than $20. I would have been better off cutting them up for firewood.

I was crushed. Standard procedure, the scale man assured me. It was industry protocol. But I knew that wood does not deteriorate that fast. Maybe in a year, yes, or two years, certainly. But in only a couple of months, the wood is still fresh and clean. But this was the policy and that was that. When I received the check, it was just enough to pay the hauler. And I resolved then and there that I would never take another load of logs to a commercial sawmill.

You see, sawmills are used to dealing with loggers who have a million dollars worth of equipment and a multi-person crew who can put together a tractor trailer load of logs a day. The rules have been written to reward them and penalize little one-man operations like mine who might spend a month putting together a load. And we little guys can't afford to haul half a dozen logs at a time—the gas prices will eat us up. So what's a little guy to do? You either join the industrial system and have a professional commercial logging operation come in and cut to avoid the custom hauling fee, or you don't sell at all, or you cut it into firewood. Or . . . or . . . or . . . you cobble together enough money to buy one of these newfangled bandsaws and circumvent the whole cotton-pickin' industry and turn your own logs into lumber.

And that's just what this forest-owner entrepreneur intended to do. Forget the hauling. Forget all the heavy machinery. Forget the logging crew. Forget the stacked deck and the discriminatory protocol. Just put the sawmill right next to the woods, cut the lumber on site, and sell valuable boards instead of raw commodity logs.

Our experience with the neighbor's mill proved the soundness of our plan. When we built a road to gain access to our forested acreage, we traded 30 acres of timber to the logger who built the road. And it is an excellent road. But the wholesale value of the logs averaged about $500 per acre. When we did our value added protocol (milling our logs into lumber) on adjacent acreage, our value per acre increased to $5,000. Now all of a sudden we could make a living from a few acres rather than the typical once-per-generation rape-cut used by most farmers. By ratcheting up the income-per-acre, we could limit our impact to just a few acres a year and let the rest continue to grow. This made economic and ecological sustainability possible.

Until the modern bandsaw revolution, on-farm mills were expensive, dangerous, energy inefficient, and labor intensive. The old-time standard circular mills used a 4-foot blade powered by a

100 horsepower Detroit diesel engine. A heavy and expensive array of cables, clamps and pigiron enabled the operator to hold a 2,000-pound log steady as it moved across the stationary 500-pound whirring blade. Many a sawyer or helper became entangled in these machines; many died gruesome deaths. "That blade cut him right in two" is the final line in many an old sawmill story.

For us, small farmers trying to make it on a shoestring, such a monstrosity was simply out of the question. At a cost of $50,000, these mills were not viable. But bandsaws were completely different. The great "aha!" occurred when one day some ingenious fellow decided it made more sense to let the log sit and just run a light bandsaw head over the stationary log rather than to have enough pig iron to hold the heavy log within 1/32 inch tolerances across a stationary blade.

The paradigm shift, for the first time, democratized sawmilling. Suddenly the whole mill could weigh just a few hundred pounds. The log just sat. A Honda engine burning two gallons of fuel a day powered the blade, pushed through manually by the operator. The blade only took 1/10th of an inch in kerf, whereas the circular mills took 1/4 inch. The power needed to remove that much sawdust is huge. Because the kerf is so small, bandsaws often recover 30 percent more lumber than circular mills. In other words, the old mills would remove an inch of wood every four cuts. A bandsaw mill only removes an inch in sawdust every ten cuts. The difference in usable lumber, called the recovery rate, is enormous.

And all of this could be had for a few thousand dollars. These mills were efficient, light, and came ready to go. Turnkey.

One other factor in the opportunity offered by the mill: local choice. With more local mills, lumber consumers could enjoy more buying options. Our community was and is no different than most: if a person wants to build something, the lumberyard is about the only game in town. And the lumberyard's stock comes from about two or three sources, all from somewhere way far away. Even if our local Home Depot sells white oak cut from this area, it wasn't kiln dried or planed here. It was exported for further processing and after a circuitous route, re-imported for sale to the locals.

This is ridiculous. Why can't a tree growing in our county be cut in our county, milled in our county, dried in our county, planed in our county, and then used in our county? Let our local economy enjoy all that value added portion, the part that creates dollar circuitry in the community rather than most of the retail dollar being siphoned off by offsite processing and transportation. To me, such reasoning is a no-brainer. But to the current industrial lumber system, it's ridiculous and virtually impossible. Imagine the numerous options available to local lumber buyers if they could visit a dozen local on-farm mills to see what struck their fancy, rather than being limited to whatever the system stocked at the local commercial lumber yard.

On many fronts, then, the bandsaw was the breakthrough we needed. The sawmill solved a multitude of problems while offering wonderful opportunities:

- Increased income per forested acre
- Profit-incentive to better steward the woodland
- Harvest plan to regenerate mature stands
- Healthier woodlot
- Low-cost lumber for building and maintenance projects
- Custom milling for neighbors
- An additional farm salary
- Additional income stream (business holon)

As I stared at the public notice in the paper, my heart sank. We were already using the neighbor's mill. We knew what it could do. We were ready to buy one for ourselves and enjoy its host of benefits. And we looked forward to offering this option to the local lumber market. But no. In the wisdom of the powers that guard the general welfare, all of this should be illegal.

This kind of policy decision-making occurs every day. Although it may start sincerely enough, the unintended consequence (if we give the policy wonks the benefit of the doubt) is to stifle innovation, creativity, and the local economy.

I decided to attend the hearing to see if I could persuade the zoning board to at least exempt sawmills operated by farmers milling logs from their own property. That seemed reasonable.

When the hearing date arrived, I cleaned up, shaved, put on a tie, left my farm work, and drove to the county building. When the amendment came before the zoning board and the chairman asked for public comment, I was the only one interested. I explained that I wanted to buy a little mill to saw my own logs from my own woods.

The board heaved a collective sigh. "We never thought of that." You can't imagine how far from reality many of these regulators actually are. Just like you and me, they live in their cocoons, too. They get up and go through their routine, read their little trade magazines, sit in their little pew at church, and become myopic just like most of us. The thought that their little amendment would cast such a broad net and scoop up a farmer like me had never even crossed their mind.

As with all industrialized paradigms, the mental picture conceived when hearing the word sawmill uttered is of mega-proportions. In our county, four sawmills operate employing 50-100 people each. Of course, they include debarking units that sell mulch to landscapers. They include giant chippers that reduce all the slabwood (soft outer cambium and bark discarded in the squaring of the log) to chips and blow it into tractor trailers that haul the material to industrial boilers. In our state, paper mills and prisons seem to be the boilers-of-choice for this material.

The words government officials hear contain a contextual component, a pre-understanding, if you will. In my experience, local commerce and small entrepreneurs never make it into their lexicon. We just don't exist. When they say the word sawmill, they aren't thinking about a $5,000, 800-lb. bandsaw powered by a 20 horsepower Honda engine. They are thinking 50 employees, tractor trailers, noise, dust and lots of infrastructure.

A few years ago when I was lobbying in the state capital for some concessions from these onerous regulations for small farmers, the senators asked the bureaucrat who headed up the state small farm agency, "What do you consider a small farm?"

179

His reply, forever etched in my memory, was simple, "Someone who only wants to sell one tractor trailer load."

My buddies and I, sitting in the audience, audibly gasped. That was as small as this government official could think. And he was supposedly the friendliest official of the friendliest agency toward small farmers. In my camp, we're talking about trunkloads of stuff, not tractor trailers. And yet the friendliest of the friendly couldn't think in terms any smaller than tractor trailer loads. That is the reality.

So it wasn't surprising that these zoning officials would have never thought that a little farmer like me might want to buy a little mill and cut lumber from the trees on his own property. "Oh, my, dopey me," seemed to be the collective response from the board.

Fortunately, they wanted to help my situation. We began wrestling with language that would describe scale. They assured me that they did not want to criminalize what I wanted to do. First, they proposed an exception with an adjective: portable. The idea being that if a mill was portable, it couldn't be but so big. However, I objected that some farmer might want to pour a concrete pad and make the mill permanent. I was also concerned about how the regulators would define portable. What if, like we have now done with our mill, a farmer towed the mill to his farm and then took the axle and wheels off? Would it still be considered portable?

In the discussion that ensued, it was clear that these officials wanted to deny my ability to cut a neighbor's log. That, to them, smacked of industrial. The compromise language we agreed on, and they subsequently voted yes to, described the exception: "except a landowner milling lumber from his own trees." I wasn't real happy with the final language, but at least I could go ahead and buy my mill.

Today, a couple of decades later, we have a wonderful bandsaw mill. It has done everything we envisioned and more. We milled the lumber for our son Daniel and daughter-in-law Sheri's house when they got married—saved $30,000. And it was better quality material. Full-cut boards, not dressed-down toothpicks.

But, every few days, we become criminals. Let me describe our criminal acts:

· A neighbor brings over some locust posts and wants them sawn to a flat edge for a board fence.

· A neighbor brings over some cedar he found and wants them cut into strips to panel around a bathtub in an old house restoration project.

· A neighbor notices a nice big walnut tree dying in his yard. He cuts it and brings the butt log over to our mill for some boards to make his wife the grandfather clock she's always wanted.

· A neighbor building a house salvages one medium-sized tree from his lot, cuts out a small log, and brings it over to mill out a mantle for the fireplace in his new house—building memories and connections.

· A friend starting a small rustic furniture business finds a farmer clearing old cedar out of a back field. The farmer agrees to let the young entrepreneur have the material for doing the work. The young fellow brings the millable trunk pieces over, runs the mill himself, and gets enough material for six bedsteads.

All of this is illegal. Why should my neighbor and friend be denied access to my mill, which sits idle 300 days a year, just because the wood didn't grow within the confines of our farm's property lines? Such a requirement is absurd. Fortunately, nobody has turned us in yet, but it could happen tomorrow.

In this same vein, the county's zoning ordinance specifically prohibits woodworking shops in agricultural zones. And slaughterhouses. Here are my questions: What better place to locate the sawmill than next to the trees? What better place to locate the woodworking shop than next to the sawmill? What better place to locate the slaughterhouse than next to the animals?

This brings us to the whole issue of zoning and integrated land use. Let's examine that in the next chapter.

Chapter 14

Zoning

The industrial model creates a Not In My Back Yard (NIMBY) approach to every processing and manufacturing activity that used to be imbedded in villages and farms.

A visit to Historic Williamsburg, and specifically the George Wythe house, brings this home in graphic detail. George Washington's Mt. Vernon, Thomas Jefferson's Monticello, James Madison's Montpelier—all of these farms enjoyed imbedded industry. In those days, these were called crafts: blacksmithing, shingle making, barrel making, carpentry, spinning, weaving, candlemaking. Complementary crafts located nearby included glass making, the cobbler, wigmaker, a couple of taverns, the gunsmith, wheelwright, tanner, and paper manufacturer.

Even the non-farmers lived with their work. Generally, the artisans lived on the second floor of the building that housed their retail/manufacturing façade on the first floor. The residential, farming, manufacturing, and processing activities were connected, integrated, imbedded in geographic proximity.

With the industrial economy, this connectedness gave way as factories and their inherent stench, noise, waste stream, and size simply made them unfriendly to nearby residences. I realize I'm close to being branded a heretic for even daring to suggest that zoning is unnecessary at best and wrong at worst, but for the sake of discussion, let's at least agree that, as currently administered, it encourages further disconnectedness in the community.

What is a farm? In our industrial agriculture, it is now merely a raw products production unit. Nothing more. And so in our enlightened state we pass "Right to Farm" laws that should really be labeled "Right to Stink Up the Neighborhood" laws. Because industrial farmers have became such irresponsible, anti-human neighbors, the corporate fraternity policy wonks protected farmers from burgeoning nuisance suits. Just like the steel mills, auto plants, and chemical factories, factory farms became unwelcome neighbors in the community. At least, unwelcome right next to peoples' homes.

The answer was to put them "out there," or at least NIMBY. But farms abutted too many neighborhoods. You can't just move a farm. As a result, farms received special location protection to continue stinking up the community, as long as they complied with "Best Management Practices." Land grant universities scrambled in the 1970s to write BMPs for all sorts of farm activities. Waste management, confinement animal housing, chemical use and disposal, logging roads, logging procedures, animal carcass disposal. Of course, BMPs always protected industrial farming models.

Then along came zoning in the mid 1960's, to further segregate people and economic sectors. Not only did this protect residential development from being located next to a farm, but it protected Burger King from being located next to DuPont. In perfect Western compartmentalized, fragmented, linear, reductionist, systematic, "it's all about me" reasoning, zoning even kept the 1,200 square foot houses away from the 2,000 square foot houses. What a social *faux pas,* that the upper middle class should rub shoulders with the lower middle class. That just wouldn't do, would it?

Amazingly, all of this was happening at the very height of the civil rights movement, when the American culture appeared to be homogenizing and becoming less segregated. Actually, the opposite was happening. By defining square footage in zones, and businesses there, and retail space there, and gift shops over there, Americans were criminalizing an integrated and imbedded local economy and its supporting commerce.

America's strength has been the freedom to use property entrepreneurially. Working where you live is a hallmark of America's history and greatness. One of the distinguishing characteristics in the American culture was this sacred right to use property as an extension of personal dreams. In other countries, the landed gentry lived in town while the peasants lived out on the land. But in America, landowners actually lived on their land, worked it, and value added its products through further processing.

The result of all this economic and land use segregation is painful and stifling. Let me present a few graphic personal examples from our own farm and our own dreams.

We would love to offer tours of our farm for schoolchildren. But we are a for-profit entity and can't just do this for free. It takes time to lead the tour, energy to run the tractor, emotional energy to teach—and hope nobody falls off the wagon and sues you. We would love to offer tours at so much per head. But if we do that, we are an educational institution, and that is prohibited in agricultural zones. A farm is not a school.

We offer a Polyface Intensive Discovery Seminar a couple of weekends every year. These are illegal. The zoning board views this as education, and farms are not educational institutions. We, however, view this as integral to our farm. Having an educated clientele, and preserving enough people doing this type of farming to patronize and thereby keep alive the support infrastructure, from implement dealers to independent chick hatcheries, is absolutely essential for us to remain viable as a farm.

If we can't buy a tractor, can't buy a chick, can't buy a hog feeder, and can't buy Genetically Modified Free grain rations from a local mill, we will cease to be viable. Unless we preserve a critical mass of purchasers, the suppliers will go out of business. All of this independent loss encourages further consolidation and centralization in agriculture. But to zoning administrators, these are completely separate issues. Remember, a farm is simply a producer of raw commodities. Period.

What if we say that we aren't educating, that this is just recreation? The education is free and we are just having people come for recreation. Oh no, in the eyes of our public officials, if anyone pays us a cent to visit the farm, we are Disneyland, and that is not acceptable in agricultural zones. We can't have theme parks sprouting up all over rural America. No way. All visitors must be free. And the officials don't care how many visitors come for free. As soon as one pays, we're a Farm-a-Disney, and that's strictly forbidden in agricultural zones.

We could have 10,000 visitors a day...as long as they don't pay. As soon as anyone pays, then were are in that kind of business—entertainment, recreation, education, whatever. And none of that is agricultural. Forget the notion that if we don't preserve the consumers' awareness of what agriculture is, perhaps tomorrow's consumers won't care about air, soil and water. After all, farmers are still the caretakers of the lion's share of these resources. And every bite of food a person eats has an effect on these resources. I'd say an awareness of these resources is critical to maintaining agriculture. But to the officials, they are totally separate ministries.

Of course, the industrial folks think they can accomplish all this awareness by hosting an annual "Farm Day" at the school. A glorified petting zoo, these interactions deny students the opportunity to see animals in their farm habitat. Come to think of it, most farmers don't want people to see animals in their farm habitat. They'd all become vegetarians. But that's another issue for another day.

Back to zoning disconnections. To show the extent of this nonsense, our county's position is that anything that leaves the farm for further processing, and then is brought back to the farm for sale, is a manufactured item and therefore illegal to sell in an agricultural zone. A neighbor has sheep. He sent the hides up to Pennsylvania for tanning and then sold them at his farm to his customers. The county shut him down on the basis that this was a manufactured item and technically made him a Wal-Mart.

Well let me get this straight. Tanning is an illegal farming activity—tanneries are prohibited. So is slaughtering, but that's another story. So I can't tan the hide on the farm, but if I send it off to a tannery, than I can't sell it when I get the finished pelt back home. That's insane. Why do I have to do the tanning? Why not use a neighbor who has the infrastructure for it and enjoys doing it? But it's the law. By this definition, when we send our beeves and hogs to the slaughterhouse, and then bring the packages back to the farm for sale, we are a Wal-Mart because we're now selling manufactured items.

We have a wonderful orchard nearby that makes the greatest cold-pressed apple juice in the world. We have a steady stream of customers who come to the farm to buy things. To service them, we have a small on-farm sales building. This orchard does not have sales traffic like we do here at our farm. We want to sell their apple juice in our on-farm store. But that is illegal because we did not produce it.

As soon as we sell the neighbor's cucumbers, or homemade salsa, or pickles, or whatever, we must have a business license. That requires the following:

- Commercial entrance. We live on a dirt road. The lane has to be widened to accommodate both entering and exiting traffic at the same time. A single lane won't do. Site distance must be maintained onto the public road, along with gravel and grading that is up to highway department standards. For crying out loud, it's not like we have 50 cars an hour coming out to the farm. If we have 20 in a day, it's a windfall.
- Handicapped parking and access. Designated areas and up-to-code everything—handrails and the works. Nobody enjoys seeing handicapped folks gain mobility more than I do, but this is my farm and my business. If I want to serve only homosexual bowlegged Vietnamese patrons, I should be able to do that. I'm not asking for government grants, taxpayer subsidies, or tax-free bonds. If I want to serve an exclusive clientele, what business is it of the government's to define who I can and can't serve and what kind of facility I want to do it in?

- Public bathrooms. Now we're into septic systems, drainage fields, hot and cold running water—thousands and thousands of dollars. Never mind that our house is 50 feet from the sales building and mom's house is 50 feet from the sales building; each one containing two bathrooms. Our personal houses don't qualify for customer restrooms.
- Up-to-code parking area. The license defines the number of parking spaces required, amount of turnaround, etc. Now we're into major excavation, graveling, and maybe even asphalting, including proper curbing, drainage, and parking barriers.
- The building must meet code. After all, we can't have a customer walking into a building that might cave in. Suddenly our home-generated lumber is suspect, because it's ungraded. For several thousand dollars, we can have a professional lumber grader come in and grade it in order to use it. If we do that, I figure we might as well just buy the outsourced stuff from Lowe's. Why complicate our life? And then we must have approved fire extinguishers, lighted exit signs. All for a building 16 ft. X 40 ft. And I'm sure our grandmother's 100-year-old wood cookstove that graces the corner and offers cozy heat would never meet code for a heating system. What if a customer's child accidentally fell against it?
- Tax on each dollar of sales. More monthly reports, higher prices. All for what? What benefit do we get out of this? For the privilege of meeting all these regulations, we get to pay thousands of dollars a year for this paperwork and regulatory oversight. What a crock.

All of this, mind you, to sell a neighbor's excess zucchini squash to a customer who is already standing in our perfectly good sales building that doesn't have any requirements as long as we only sell what we produce. Folks can park in potholes, walk into a snake-infested shack, and take a leak behind the grapevine if they buy our tomatoes, our chicken, and our lumber. But if we offer one cucumber from the neighbor's garden, we must spend perhaps hundreds of thousands of dollars in upgrades and comply with a host of regulations.

Think about it a minute. Wouldn't it be wonderful if our neighbors could tap into our sales building, and if our customers, who have invested time and money to connect with their food by coming to the country, could access our neighbors' artisanal products? That would be win-win for all of us. It would more fully utilize the local production, the infrastructure, and the transportation investment. By leveraging what everyone is already doing, the synergy would make the whole worth more than the sum of the parts. Why duplicate effort? Why require every farm to have a sales building, complete with scales, cash registers, and cashiers, when a centrally located one serving a dozen farmers would make the whole ball of wax efficient and economical? Because that would make too much sense. And it might actually compete with WalMart.

Many times, as travelers approach rural counties like ours, they will see a sign on the interstate at the jurisidictional boundary proclaiming: "County XYZ: A Business Friendly Location." Why don't they ever erect a sign saying: "County XYZ: A Farmer Friendly Location?" The reason is because farms are now viewed as liabilities rather than assets. And farmland just gets in the way of progress like another Wal-Mart, housing development, or strip mall. The local big wigs don't want to rule over a bunch of farm peasants; their portfolios and prestige only accelerate when they rule over more buildings and industries.

The result of all this nonsense is that it divorces the farm from everything that used to be done there, processing, marketing, education, recreation. And when the farmer divorces everything farther up the food system, he bears no responsibility. And on a grander scale, when a society segregates itself, the consequences affect the economy, the emotions, and the ecology. That's one reason why it's easy for pro-lifers to eat factory-raised animals that disrespect everything sacred about creation. And that is why it's easy for rabid environmentalists to hate chainsaws even though they snuggle into a mattress supported by a black walnut bedstead.

In fact, writing this book on my own farm at my own desk with my own computer is illegal—it's not permitted in an agricultural zone without a special use permit. Writing is not an agricultural pursuit. Oh, let me clarify that a bit. If I did it for free, it's okay. But if I sell one essay, then I'm illegal without a government permit. As with so many of these activities, clearly the problem is not the activity itself; it's the commerce of the activity that's the problem. But what is it about the commerce of the activity that suddenly makes it a threat to the neighbors? They certainly can't tell which essays I write for free and which ones for sale.

I can sit here and generate a hundred letters to the editor, email them to the newspaper, and be perfectly legal. But if I email one to a magazine and receive a check in the mail, which comes along with a handful of other items to my mailbox anyway, then suddenly this is a threat to my neighboring farmers.

The problem is that when we take a holistic view, we can't separate the farm from all these other elements. If I had never offered salad bar beef to my neighbors, we would have been stuck in the commodity business. If we had been stuck in the commodity business, we would not have been able to make a living on this small farm. If we had not been able to make a living on this small farm, we would not have had a viable agricultural enterprise. Had we not written about a viable enterprise, it would not have grown enough to add our children's salaries. If it could not have grown, it would have withered before it started, and it wouldn't exist.

And therein lies the crux of the problem: it wouldn't exist. We cannot know how many farms and how much good, locally produced food is unavailable because of these regulations. And that is the ultimate tragedy, that much good simply cannot exist. In discussing the divine authority of government, Paul in Romans 13:3-4 says, "For rulers are not a terror to good works, but to the evil...do that which is good, and you will have praise of the same; For he [the government agent] is the minister of God to you for good. But if you do that which is evil, be afraid; for he bears not the sword in vain: for he is the minister of God, a revenger to execute wrath upon him that does evil."

Clearly, God intends government agents to discourage evil and encourage good. Let's just list a few of the things enumerated in the preceding examples as being evil:

- Hosting kindergartners for $1 a head to get a hayride tour of the farm, pet the chicks, see the cows, and handle compost.
- Conducting a seminar for paying students to learn how to farm in a way that is ecologically, emotionally, and economically sound.
- Offering a $100 per plate dinner in our living room of on-farm produced and processed foods to patrons yearning for a delightful evening out.
- Writing an essay for a magazine that will pay for the article.
- Selling a T-bone steak that was processed in a nearby federally inspected abattoir.
- Selling a neighbor's extra pumpkins in the fall.
- Milling a neighbor's log into a mantle for the fireplace in his new house.

Are these activities evil? I challenge anyone to explain to me how these are evil. Not only are they not evil, they are decidedly wholesome, good, neighbor-friendly things to do. These are the activities that build communities and connect neighbors. Doesn't enough evil exist in the world already without our culture demonizing and criminalizing all sorts of good behavior? Why must we add to the legitimate pool of evil those activities that are the heartbeat of a community's fabric?

I return now to the original question: What is a farm? The result of all this criminalization is that the farm has been relegated to colonial status. In today's world, a farm can only produce raw commodities to feed the global industrial complex. This is simply a permutation of the colonial theme. How did Spain view the New World? How did England view the American colonies? How did the British Empire view the South Pacific Islands? How did France view Vietnam? How does the United States view the oil-rich countries of the Middle East? How do Tyson and Pilgrim's Pride view the poultry farmer?

In all these cases, the imperialist—whether it's a government or a multi-national corporation or an entity aspiring to be an empire—views the controlled people or lands as raw materials producers for the manufacturing machine. And in each case, the empire cannot allow the colonist to tap into the value added components, lest the most profitable elements be siphoned off by upstart competition. The colonist, then, is relegated to peasanthood, a feudal serf in the great machinery of the industrial global complex. And when he mines out the last of his raw materials, the lords of manufacturing leave him to court the next colonist willing to work for pauper's wages and sell raw commodities until they are used up.

The elitism of the industrial and regulatory cartel in the food system is obvious to any who look at the flow of power and money. The rules of the game are such that players consistently move their wealth and power to the corporate lords and the government nobles. This wealth flow continues unabated, and calling in government agents to regulate the use of the word organic played right into the rules of the game. Why would farmers who defined organic, created the movement, and developed cultural awareness voluntarily give up all their ownership, all their hard-won credentials, to a nameless, faceless government agent?

The so-called farm crisis is really not a crisis at all. It is simply a collective willingness on the part of farmers to be colonists, to be complicit peasants. For the most part, unfortunately, farmers do not aspire to do the kinds of things enumerated here as being illegal, and rather resent that a handful like me have alternative dreams. The American Farm Bureau Federation and the National Cattleman Beef Association are two prominent alleged farm organizations that come to mind who are lackeys for the empire-builders. The policies these organizations advocate insure continued impoverishment for the farmers and rural wealth flow toward the financial and trading centers.

Of course, as the raw commodity price drops, farms must get bigger and cut more corners. This is why confinement buildings keep getting bigger. This is why cornfields keep getting bigger. This is why we now burn cow manure to generate electricity—get rid of it as cheaply as possible. And with all this change comes more odor, more assaults against nature, and more diseases.

In all the decades I have been advocating for freedom and food choice, these organizations vilify my message and they collectively paint me as an enemy of the farmer. Indeed, an enemy of the food system, but that's a discussion for another chapter. The point is that industrial farmers are as duplicitous in this empire-colonist scheme as non-farmers. Neither supermarket shoppers nor industrial farmers have a clue what would be available in their communities were these activities legalized.

This is all the natural extension of a totally fragmented, reductionist, compartmentalized mentality. The problems are somewhere over there; someone else's responsibility. The relationship between my materialism and the energy crunch, between my menu and the water I drink is completely severed. In today's world, for the first time in human history, a person can move into a neighborhood, build a house, and live, and never consider the climate or ecology of the surroundings. He connects to a water pipe coming in, a sewer pipe going out, buys food at the corner supermarket, clothes at the strip mall, and energy from a robot-answering phone machine utility.

Throughout human history, when a person moved into an area, he had to think about where the water came from. He had to think about where his food came from. He had to think about how he was going to heat his house. He had to think about what he would do with trash and excrement. He had to think about where his shelter materials would come from. In the modern era, a person can build his house from outsourced materials shipped in from across the world, plug his TV into an outlet that delivers power from a source somewhere far away, plumb his sink into a pipe, affix his toilet to a pipe, and buy food from the supermarket that gets packages from who knows where.

No knowledge, no responsibility, no connections. And therefore no understanding of the relationships, the intricate natural checks and balances economically and ecologically, that are required to sustain the house, energy, air, water, and food. The people can play their Nintendo, open their plastic packages, send their dues to Sierra Clubs for their environmental penance, and vote for politicians who vote daily for policies that destroy local farms. If I could write for them, entertain them, educate them, feed them, and connect with them, what a wonderful thing that would be. But at every attempted connection, a host of bureaucrats stands in the way, the power of the state backing them up, to call such connections evil.

So what is a farm? Certainly if Hillary Clinton's notion that It Takes a Village to Raise a Child has any credence, I would suggest that it takes a community to preserve a farm. To divorce a farm from its neighbors, its customers, the body of knowledge regarding ecological land stewardship and earthworm activities, is to destroy the farm. It cannot exist separated from the rich cultural soup that sustains it. It is not just a raw producer of commodities.

A farm includes the passion of the farmer's heart, the interest of the farm's customers, the biological activity in the soil, the pleasantness of the air above the farm—it's everything touching, emanating from, and supplying that piece of landscape. A farm is virtually a living organism. The tragedy of our time is that cultural philosophies and market realities are squeezing life's vitality out of most farms. And that is why the average farmer is now 60 years old. Serfdom just doesn't attract the best and brightest.

A farm relegated to production of raw commodities is not a farm at all. It is a temporary blip until the land is used up, the water polluted, the neighbors nauseated, and the air unbreathable. The farmhouse, the concrete, the machinery, and outbuildings become relics of a bygone vibrancy when another family farm moves to the city financial centers for relief.

The information economy is allowing thousands of people to move out of urban centers and take up residence in the countryside. Slowly the commute to the office is becoming unnecessary. The home office, telecommuting, and being satellite uplinked are new business realities describing these trends. For decades farmers have been trying to figure out how to get their stuff to the customers. And now the customers are coming to rural

America. As a culture, we have a new opportunity to imbed farms into the residential landscape.

But with current zoning policies, we as a people insure that these new neighbors will not set foot on the farm. That indeed the farms will not be imbedded, but will be over there somewhere, out of sight and out of mind. And that the children who could grow up knowing their local farmer will rather grow up knowing only the supermarket and its artificial food climate. Segregation wasn't good for America's races, and it isn't good for our food system. Integration can only occur when we realize that a farm is far more than the producer of raw commodities.

Perhaps we need a "Right to be Imbedded in the Community" law to preserve the integrated farm. As it currently stands, most of the activities required to preserve the kind of farm yearned for by most people is simply illegal. And that is evil, not righteous.

Chapter 15

Labor

O f all the pressing issues farmers face, probably the most critical one is labor. The stereotypical large farm family did not occur because farmers had nothing better to do with their time, but because farms are notoriously shorthanded. The imperative to grow your own labor is very real in farm country.

Dad read an article in the 1960s that prophesied that the farm of the 1980s would be operated from a push-button console. The farmer's workday would commence at the console, which looked much like the flight deck of a Boeing 777 and it ended in the evening at that console. Dad lived until 1988, and during the early part of the decade, when we were working at some especially laborious task like trying to pry a huge rock out of a post hole, he would look at me, sweat dripping off both our faces, and exult with a wry grin: "Here's some more of that push button farmin'!"

It was a precious shared joke between us. Of course, push button farmin' will never come because real food requires real plants and real animals which need real caretakers to love and nurture them. Real caretakers are real people. Just wanted to make sure we were all on the same page. Be assured that the industrial sector is still preaching the virtues and practicality of farmerless farms.

Even with all the modern machinery and clever inventions, our farm still has plenty of labor needs. And our community is full of teenagers who desperately need meaningful tasks to occupy their time and make them feel important and necessary. Just for the record, here is a list of things I believe teens 16 or under can do on our farm:

- Mow the lawn.
- Drive the 4-wheeler to the Eggmobile and Feathernet and gather eggs.
- Kill chickens.
- Scald chickens.
- Pull off heads and cut off feet of chickens.
- Eviscerate chickens.
- De-pinfeather chickens, aided by a dull knife.
- Split wood.
- Build things using a cordless screwdriver and circle saw.
- Drive the tractor towing an empty hay wagon to the field and tow a full one back to the barn.
- Drive the tractor to rake hay.
- Drive the pickup out to get a load of firewood or a load of anything.
- Use the grinder in the shop to take the burr off a weld.
- Weld.
- Cut steel with an acetylene torch.
- Prune trees with a handsaw.
- Put bales of hay onto a hay elevator.
- Operate a generator.
- Operate a water pump.
- Till the garden with a rototiller.
- Drill a hole in a board.
- Cut a board on the table saw.

But all of this is illegal because it involves a power tool or motor, and no one 16 or under can legally operate those things for pay unless they are employed by their parents. Now I want us all to take a deep breath and think about this a minute.

A 16-year-old can get behind the wheel of an automobile and drive it 70 miles an hour down the freeway, but can't drive a 4-wheeler out to the back field and gather eggs. Or operate a cordless screwdriver. Does that strike you as odd?

Our son, Daniel and daughter, Rachel, were doing all these things at 10—and responsibly. With increasing regularity, the newspaper carries stories about teenagers involved in gang activities, roving the streets at 2 a.m. getting into trouble. The social workers have a single refrain in every story: "Oh, if our community would just provide something for them to do!" The incessant wailing is pitiful.

When our kids were 14 and 15, not to mention 18, they weren't out roving the streets at 2 a.m. They went to bed at 10 p.m. because they were tired from a day of meaningful work. Modern society is wringing its collective hands, wondering what to do to "get these kids off the streets." And teens desperately need to feel necessary. But when we've outlawed everything they used to do at that age to prepare them for a meaningful vocational existence, all they can legally do is play computer games.

Of course, some teens do get jobs. But the stigma against a farm job is deep. I remember well when I was in high school and told the guidance counselor I wanted to be a farmer. She said it was a waste of my life. Practically in tears she begged me to do something more valuable than to be a farmer.

Not long ago I was judging a 4-H public speaking contest, and one of the contestants did a speech about his run-in with his school guidance counselor. His experience was the same as mine. School officials conspire against smart kids to make sure they don't squander their education by being farmers. And of course they say the word farmer as if to even speak the word is beneath their dignity. It's spoken with a kind of slobbery spittal aspiration. A tone of exasperation and condescension.

But just for a minute, on a whim, let's imagine that a 15-year-old neighbor boy does want to come over and help dress chickens. Or drive the 4-wheeler out to the field and gather eggs. Or help build a hay wagon deck. If his parents are happy with the job description, he's happy with it, and we're happy with it, why is this activity suddenly a government responsibility?

A young person old enough to be squired into a classroom and instructed in the intricate uses of a condom can't operate a cordless screwdriver? Am I missing something here? Teens who can take a hunter safety course and be state-licensed to carry a high-powered rifle into the woods to shoot a bear can't push the chicken scalder button or cut off chicken feet? What have we done to ourselves as a society?

We've taken away the dignity of early adulthood and replaced it with foolishness and recreation. Meanwhile, the farmers in the community are denied the benefits of all this untapped youthful zest. It's a tragedy of epic proportions, and in my opinion a massive cultural *faux pas*. At the risk of sounding like Scrooge, I think child labor laws should never have been passed.

Cultures cycle. Social norms do not stay the same over time; they move with information and new realities. Just like after Upton Sinclair wrote <u>The Jungle</u> in 1906 and beef consumption dropped nearly 50 percent in less than six months. That was a massive cultural shift in response to new information.

The current consumer demand for grass-based meat and poultry is a direct response to nutritional information. Things change. Look at the reshuffling of the official USDA food pyramid. Look at how the culture responded to the Vietnam war. Or for that matter, the Iraq war. Look at the recycling movement. Phonics to Dick-and-Jane, then back to phonics.

Once the majority has the votes to ban something, it's probably already on the way out. I wear a seat belt not because it's the law, but because it's safer. I wonder if the rebellious fringe who refuses to wear seat belts as an act of defiance against being required to is any bigger than the number of folks who would not wear it only out of complacency or negligence had mandatory belt laws never been passed. I guess we will never know.

As these types of laws proliferate, all of us find fewer and fewer spots of autonomy left. Being able to make self-directed decisions is critical for expressing our humanness. Not that every individual expression is okay—i.e. violations of the Ten Commandments—but these basic moral codes are a far cry from the kind of micro-behavioral codes emanating from today's politicians. The Romans had a saying that the better the government, the fewer the laws.

The reason I would have opposed the first child labor laws is the same reason I oppose today's permutations. A willing worker and a willing employer should be able to come to an agreement without governmental intervention. I would certainly oppose an employer's ability to conscript, by coercion, any employee regardless of age. But maturity comes to different people differently, as do abilities and interests. Why doesn't anyone jump up and down and yell "Exploitation!" when a child prodigy plays beautifully on a musical instrument?

But let an equal-aged child demonstrate an incredible aptitude to milk a cow or eviscerate a chicken, and the culture assumes exploitation. In a culture where farm work is philosophically assigned to the slow and inept, bright children doing these tasks is assumed to be abusive. But those of us who are familiar with children who grew up doing farm chores can attest that they tend to grow up with purpose, self-respect, and appreciative of others' efforts. We are raising a self-absorbed generation precisely because they have been denied the necessary education in human activity; namely, serving others through our abilities and appreciating all the members of the team. I would go so far as to say the current outlawed youthful activities amount to cultural child abuse that is increasingly coming to haunt us as a society.

The fact that, as a culture, we began outlawing child workplace exploitation is proof itself that the culture was changing. Unfortunately, people are impatient and too quickly want to use government to coerce onto the rest of society the supposed enlightenment that they have already embraced. The religious right, viewing their position as righteous, quickly seeks to use the long arm of the government to right their perceived societal wrongs.

And the pagans, humanists, or liberals—whatever you want to call them—do the same thing in elements like child labor laws, increasing the minimum wage, and mandating seat belts. What is wrong with letting culture change and move at its own pace? Why must either side quickly invoke governmental force to speed its agenda? At the risk of truly being branded an extremist, I'll wade into even the civil rights movement, which was moving apace, steadily and surely, through the efforts of charismatics like Martin Luther King.

But as soon as the government became involved with the process, we ended up forcing it and causing more harm than good. Does anyone really believe we have better race relations in America today than we would have had had the steady progress achieved during the early 1960s been allowed to move no faster than society could metabolize it? Instead, we have reverse discrimination, forced school busing, and as much if not more racial hatred today than in 1965. Every movement has its fringes, and the Ku Klux Klan certainly did not represent the average person during the civil rights awakening. Reverse discrimination now fosters resentment that never would have occurred had the movement been allowed to move forward without federal government heavy-handedness.

I have a friend who at 13 decided he wanted to cut firewood for a living. His family had an old 1965 International loadstar dump truck with a two-speed rear axle. A little big for his age, he began cutting firewood—yes, with a chain saw. He could drive that truck as well as any adult, and delivered a load to a customer in Waynesboro about 15 miles away.

A neighbor saw the load being dumped and walked over to ask for one himself. The boy said he'd be there the following Saturday with the load. He cut the wood after school and in a week had another dump truck load. He arrived at the man's house on schedule. The customer was dubious about the boy's ability to back the truck around a crooked driveway, swing set, picnic area, and dump it onto a concrete pad in the corner of the back yard.

The boy assured the customer that it was no problem and proceeded to back the truck expertly right up to the pad. He dumped the load and pulled out. What I didn't tell you was that this man happened to be a city policeman, and this week happened to be in uniform because he was getting ready to go to work. When he got ready to pay the boy, he asked him: "Son, how old are you?"

My friend, who confesses that his heart skipped a beat when he pulled in and realized he was delivering wood to a policeman's house, just looked down from the cab and confidently but with a mischievous grin responded, "How old do you want me to be?"

Taken aback, the policeman asked him, "Do you have a license?"

To which the 13-year-old replied, "I have a couple, but I'm not sure I have the one you're interested in."

Completely enthralled by the whimsical youngster, the policeman said, "Listen, son, I'll follow you to the edge of the city limits. And please, please, don't drive that truck back into this city."

Two weeks later, the boy had another load to deliver to Waynesboro. As luck would have it, when he entered the city, there on routine traffic patrol was his firewood customer friend. They locked eyes, and the policeman just shook his head, took off his hat, and covered his eyes. Now there is a true public servant. That boy could drive that truck as expertly as any adult. How could anyone deny this self-motivated, bright-eyed and bushy-tailed boy his firewood business? To be honest with you, I wouldn't let my 13-year-old do this. And it probably wasn't the wisest thing to do.

But for crying out loud, must we criminalize everything we don't think is wise? I don't think it's wise to pay credit card interest, but plenty of people do. I don't think it's wise to smoke, but people do. I don't think it's wise to buy lottery tickets or gamble in Las Vegas or climb Mt. Everest. But do we want a society that outlaws everything that 51 percent of the people think is unwise? If the truth be told, most of these regulations occur without involving 51 percent of the people. Most occur because a few influential people buy off or cozy up to the politicians or bureaucrats who act like marionettes on the end of a string. Every time I become involved in the political process, I'm amazed at how few people are actually involved. A vocal few, or wealthy few, can influence major cultural changes without 95 percent of the people having a clue what just happened. Anyway, let's get back to labor.

With the homeschooling movement, many teens are finishing school at 16 or before and are ready to move forward with their lives. Some of them apply for an apprenticeship here at our farm because they want to be full-time environmentally-friendly farmers. How devastating it is for us to have to tell them we can't legally take them.

These young people are emotionally and physically mature enough for a driver's license, but they can't push our lawnmower. The problem with invoking government force to correct every little thing that a culture has determined to be wrong is that once the bureaucracy is in place, the correction always becomes overcorrection. To justify their existence, the government agents cannot be content to solve the initial problem. They continue to push and push beyond the point of reason until the original agenda becomes mired in absurdity.

If a 16-year-old wants to come here and work, willingly, and we are happy to have him, who is being harmed by his apprenticeship? Why must he wait and spin his wheels for two years? If he comes here and gets the learning experience earlier, he can start down his entrepreneurial way earlier. And don't give me this "we must protect him from himself" business.

If that really were a duty of government, we could justify putting everyone in straightjackets to protect us from ourselves. I might jab myself in the eye with my toothbrush in the morning. I think we should legislate toothbrushing licenses before someone can use one. I might burn myself on the coffeemaker in the morning. Better issue a coffeemaker license while we're at it. And don't get me started on drinking.

Now let's assume that we have a legal apprentice, aged 18. How do we figure out a legal way to pay him? According to minimum wage laws, he must be paid per hour, plus overtime. But this is an educational experience. At least, that's what we think. But unless it's a bona fide credentialed educational institution of sorts, it's just an exploitive working situation as far as the government is concerned.

In the winter, we work about 20 hours a week. In the summer, we work 80 hours a week. One is as important a part of the experience as the other. But according to the government, we can't make up for 80-hour weeks in the summer with lots of free time and reading time in the winter. All the hours must be corrected within the two week legal limit.

Remember, all of this has been presented beforehand, and the apprentice voluntarily—indeed, eagerly—signed up for it. Plenty of farmers have been taken to the cleaners for violating labor laws in this regard. The government views what we do as exploitive of labor. But if the laborer happily agrees to these conditions, how can it be exploitive? Oh, that's right. How silly of me. The worker is too stupid to know when he's being exploited. Workers happily walk to their own gallows.

— That's another principle of government agents. They assume that everyone but them is a dolt. An ignorant, nonthinking dupe. Does it ever occur to these bureaucrats—not to mention the rest of society that feels compelled to rescue laborers from themselves—that maybe some people enjoy doing what they do? Even if what they do wouldn't make me happy? We are quick to inject our own standards of happiness and satisfaction on others when we often don't have a clue what's going on in their heads.

I can't imagine being a sign holder for a highway construction crew—you know, the ones with SLOW on one side and STOP on the other. You couldn't pay me enough to do that job. And yet I'm confident that many of them are absolutely satisfied and happy doing it. And even if they aren't happy, is that my business? Since when is it the government's business to insure that employees are happy? These sign holders smile and wave. Who knows? Maybe they get a power rush every time they turn that sign to authorize the stopping or starting of several tons of pig iron and human flesh. And as far as I'm concerned, if an employer can find someone willing to do that job for 50 cents an hour, that should be his prerogative. Maybe some retired person just wants a feeling of power for a day and doesn't want to earn anything. Why is that the business of you or I to invoke the government's force into that voluntary relationship?

The reason this is important is because a local food system requires local farmers to access local labor. Part of a viable local food system is a local infrastructure, which includes the labor. One of the characteristics of non-local food systems is an inherent need to find cheap labor from outside the community. This influx of cheap labor can drastically change the social structure of a neighborhood. The resultant drastic increase in English-as-a-Second-Language classrooms and instructors, police protection, and translators at the courthouse can literally convulse and traumatize a community. This is not xenophobia; it is responsible social awareness.

A local food system that is socially responsible should be able to hire local labor. And farms have always depended on young energy or lower cost labor. Farm communities had indigenous floating farmhands. They would work a few days on one place, then a few days somewhere else. Gradually, these folks have gotten older or moved to the city or gone on welfare.

Rural communities no longer have this bank of farmhand workers. Where will we get our labor? I would rather use apprentices who want to come here and learn while they work than I would import illegal aliens. To me, that seems more socially responsible for my community. My first desire would be to hire people in the community, but too few of them are interested in this kind of work. The next best thing is to find folks who are compatible with our community so that while they are here, they don't create a spasm in the neighborhood.

And they come, happily. In fact, we turn down more than we accept. Since this is an intimate and immersive learning experience, we want them to stay on the farm. Oh boy, here we go again. Where will they stay?

According to government housing standards, the accommodations must be of a certain type—generally far too expensive for a small farm to afford. If we pay them enough to stay in a rental place in town, the whole apprenticeship idea is defeated. We end up just paying a bunch of unnecessary money for gas and they waste their time in travel. The most effective way is to live here on the farm.

But our county doesn't want housing on farms. Zoning has deemed housing incompatible with farms. Housing is for the farmer. Period. Farms are for pretty sightseeing, for bicyclists from the city to come out and enjoy. They are for scenery and recreation. We don't want extra houses and we certainly don't want extra people.

I know farmers who have a waiting list of young apprentices ready to come for a summer and live in tents or a modified chicken house to enjoy a farm experience. And some of these young people go on to become extremely successful farmers. If we are going to build a local food system, we must offer mentorship opportunities that are easy for the farmer to institute. If before taking apprentices I need to go through a special use permit hearing, and very likely lose, and then build a $100,000 up-to-code house, I'll never begin the apprenticeship program.

Then aspiring young people who want to take this mantle and become local food producers never have the opportunity to experience real life on a real farm serving real neighbors. Instead, they go into some Wall Street job and become miserable 45-year-olds who desperately want to escape their Dilbert cubicles at the end of the expressway and do something meaningful with their teenage children before their children are doomed to the same cog-in-a-globalist's-wheel life. How will these marvelous young people have this opportunity when society via the government has decreed that they may not camp in tents for the summer?

Maybe these young people think this is a lark, a far more wonderful experience than living under asphalt shingles and looking at drywall? What business is it of mine, yours, or anyone else's what these consenting young people find fulfilling? I find it absurd that in our culture school dispensaries can give free condoms to consenting teens, but those same teens can't make their own decision to live in a tent for a summer on a farm to learn meaningful food production. In typical do-gooder intervention, the bureaucrat assumes the farmer is exploiting these people, abusing them, and offering substandard accommodations. Essentially this is viewed as modern day slavery.

Only one big difference. These apprentices chose this arrangement. They aren't shackled to anything. On our farm, we don't even sign any contracts. You're free to go and we're free to kick you out. It's a total transparent, open-ended handshake deal—a deal, I submit, that is nobody else's business.

A couple of decades ago I was driving home one afternoon from a whirlwind around-the-state bull-seeking trip. Yes, I was shopping for a bull. Not just bull, but A bull. I picked up a hitchhiker who turned out to be a homeless fellow. He was no dummy, I assure you of that. I asked him where he was going and he asked me where I was going. When I told him there was a good shelter in Staunton, our nearest city, he said he'd be glad to go there.

Of course he had his down-and-out story and I listened attentively. I admit to being gullible about these things. It's a weakness. I would rather be taken advantage of and be accused of being undiscerning but soft hearted, than be hard nosed and be accused of being hard hearted. We were in the car for more than an hour and covered a lot of territory in our conversation by the time I dropped him off at the homeless shelter.

To set the context, this was before Teresa and I were famous. I was driving a $500 car; we were living in the attic of the farm house; our gross annual income was about $7,000 a year. We were making it, but we had to really watch the pennies. We had a big garden, a milk cow, and our own firewood. I always said that if we could figure out how to grow our own toilet paper and Kleenex, we could just about pull the plug on society.

I told him that if he ever needed anything, to call me. And I meant it. That was on a Saturday. The following Monday, he called me and asked if I could use him. I assured him that I could not afford to pay him much, but if he wanted to stick around for awhile and get his life together, I'd try to help. He wanted to do that. Since he had no transportation, I drove in the next morning and picked him up.

He was a very capable worker. He ate lunch with us and was wonderful with the children, who were very small at the time. I paid him $20 and took him back. He said he wanted to come back the next day. This went on for a week. He quit smoking during that week and it was clear he really liked our stable family life and appreciated being there with us. We enjoyed him as well.

The shelter had a policy not to let people stay longer than a couple of weeks unless they were gainfully employed and seemed to be making progress getting their life together. His time was coming to a close so he had the normal evaluation interview with the mission administrator. She told him that the reason he was in this predicament was because of rich farmers like me who exploited people like him. He was devastated, and asked me if what he was doing was real work or just unnecessary drivel.

I assured him that he was being a great help, but since I was living on a little more than $100 a week, it was hard for me to pay him $100 a week. Things were tight. But we worked out a plan that we could renovate an old chicken house on the property, put in a woodstove, and provide him a place to live for the winter. He loved the idea. His eyes brimmed with tears at the thought of a stable place for that long.

We continued this program for another week. It cramped us to make the 45-minute round trip twice a day, but we could see progress in him and enjoyed being this personally involved in a work of charity. The following Monday my brother and his family were coming for two weeks (they were missionaries with New Tribes Mission in Indonesia) and we knew things would not be normal. So I contacted a friend and asked if he could use William (not his real name) for two weeks until our situation settled down.

He said he could so all seemed in good order. It was a huge undertaking at that time in our lives. Teresa and I had to deal with our qualms about a vagabond living that intimately with us with our two young children, but we decided that since we were there all the time and the children were never out of our sight, we would commit ourselves to doing our best to help him.

My friend, who operated a nursery business and employed some 30 people, arranged for one of his employees to pick William up on Monday. William didn't meet the ride. He didn't meet the ride on Tuesday. I called my friend Tuesday night and asked him how William was working out. "He never showed up," was the response.

I immediately called William at the mission. He wasn't there. I called the next morning and got him. "What's the deal?" I asked him. By this time, having worked together for two weeks here on the farm, we were over the niceties of tip-toeing conversation.

"Oh, I fell and hurt my leg and had to go to the emergency room..." He had a whole sob story. Somehow, though, it didn't make sense. I called the hospital emergency room and asked if William had been there. No record.

I called William back, "William, I've got a problem. I helped you for two weeks and was proud of what you were doing. I got a real job for you for two weeks while I'm dealing with family stuff. The hospital says you weren't there. Tell me the real story."

He hung up, skipped town, and we've never heard from him since. Now I'm sure you Sean Hannity disciples are saying, "Joel, you really are gullible. Don't you know all these homeless folks are social trash? You can't help them. Forget it."

And you liberals are saying, "Why didn't you pay him more? You cheapskate. You exploitive businessman. He skipped town because he was tired of being taken advantage of."

I don't know which of these responses I detest more. I don't even know if either or both of them are right. But I do know this: as honest as I can be, we put ourselves out; we did our best; we believed—and still do—that we were doing the right thing as well as we knew how. If he wasn't happy, he sure didn't express that to us. And yes, he probably was just a drifter. Maybe the thought of actually being responsible and settling down scared him. Maybe he got drunk and fell in with the wrong crowd Sunday night and just couldn't bring himself to face me ever again because he felt like of all the people he'd met recently, I really cared.

To this day, lots of this story remains unanswered. But I do know this. For that administrator to say I was a wealthy predator on society's dregs was equivalent to calling charity evil. She didn't know the situation and had no right to say that. And the other thing is that our relationship was unique. Our needs were unique. Our connection was unique. What we appreciated in each other was unique.

With the trouble he was having on booze, cigarettes, and gambling, he didn't need $100 a week. He needed love and stability. Throwing more money at people who don't know how to spend it is not what they need. I was prepared to make a budget with William, to help him discover independence. And for the government or any one else to say that I took advantage of him by not providing better housing, better pay, or whatever is just plain wrong.

When a person can't even do good without such benevolence being demonized by fat paycheck bureaucrats, milk-fed politicians, and platitudinous social activists, then we're in a sad state of affairs. A willing worker of any stripe and a willing employer should be able to come to a consensual arrangement without any government agent deeming the transaction unfair at best or exploitive at worst.

That I have a dim view of government welfare is quite obvious. I went up to a fellow holding a sign at a supermarket: "Will work for food." I offered him to come out to the farm and work, in exchange for a great meal. He wasn't interested. He wanted a handout.

Several years ago I received a call from the regional area Food Bank. The fellow said they needed someone to take away unused food. They had a local pig farmer who had been coming to get stuff and he died suddenly. They were looking for someone who had pigs who could use the excess food.

They said they tried putting it in the dumpster, but it just wasn't good for donations when people saw them dumping food in dumpsters. It's hard to make the case for needing more food when the food bank begins throwing food in dumpsters. We began going over there in the pickup and bringing home food. Wow! What a lesson in the American hunger racket.

I had never been in one of these places before, and when I walked in I was struck by the junk food. Mountains and mountains of junk food. Anyway, we began going weekly and bringing home heads of lettuce, buckets and buckets of eggs, bread and pastries. One week they put on two tons—yes, you read that right, 4,000 pounds—of perfectly good sweet potatoes. Now dear folks, I love sweet potatoes. Sweet potato casserole, sweet potato pie—it just doesn't get any better.

I asked the food bank staff: "What's with all these sweet potatoes? These are perfectly good." I was aghast at such a waste.

"Sweet potatoes are poor man's food. Our people don't want poor man's food. They want processed foods—that's rich man's food."

I couldn't believe my ears. For the next month, our family enjoyed the best sweet potatoes you could ever want. We ate and ate and ate, but we just couldn't eat through those thousands of pounds, so we still ended up feeding most of them to the pigs. But it sure taught me a lesson about hunger in this country.

We finally quit when we began routinely bringing home mountains of ready-to-bake biscuits in those cardboard cans. I had never even seen these things before. I didn't know they needed to be refrigerated. We'd unload boxes and boxes of them in the shed and then for the next several days, until we could feed them to the pigs, walk by the stack to the melody of exploding cans. The trash. Oh, the trash. Barrels and barrels of trash.

That helped us understand the true goal of America's food system: see how much packaging can be amassed around each calorie of food in order to keep the garbage collectors busy and the landfills busy. Packaging alone is outrageous in America's food system. And to top it off, the pork from our hogs deteriorated in quality due to the poor Food Bank diet. The meat became soft and tasteless.

I've had numerous preppy college co-eds come out for farm tours, and they literally can't walk half a mile. The co-eds from working-class colleges can handle the walking tour no problem. But the preppier the college, the more out of shape they are. They must be eating food bank fare. Finally, we got tired of the poor quality of the food bank food—it wasn't good enough for the pigs. And we got tired of barrels and barrels of packaging trash. After a couple of months, we quit getting food bank extras.

Labor, food, and farming. What should be a beautiful dance is a tangled web of propaganda, deceit, and demonizing. Every time I read the official take by lobbyists and big organizations, I'm reminded how easy it is for any and all of us to be taken in by snake oil salesmen. But instead of outlawing the salesman, and the freedom we have to own our decisions and our bodies, let's grow up. Let's mature. Let's research and dig out the truth. And I point the finger at myself as much as anyone else. Finding the truth is a constant challenge. It doesn't come easily, but it's worth finding.

Chapter 16

Housing

"It's illegal to build a house less than 900 square feet," said the building inspector to our son Daniel. Daniel and I looked at each other and realized this multi-generational farming plan would not be as easy as we had hoped. Even in our own desire for a simple, inexpensive house, what we wanted to do was illegal.

Perhaps few things are more important in preserving local farms than creating a climate that allows a seamless transition between generations. Just like any business, continuity must be maintained to keep the door open when the previous generation can't handle the heavy lifting any more. Because farming involves such intimate knowledge of the land and everything surrounding that land, the most efficient way to preserve the business is to grow replacement stewards from within.

Traditionally the most successful farms have been those that enjoyed one or more children taking over the reins as Mom and Dad aged. From an accumulation of tribal or familial wealth standpoint, this is the historical technique. Elderly wisdom leveraged on youthful energy is the way to accomplish the most correct work the fastest. Correctness is determined by elderly wisdom, and speed is determined by youthful energy.

In our own case, Dad and Mom worked off-farm jobs to pay for it. When Teresa and I came along, the other siblings were already gone from home and Mom and Dad were rattling around in the big old farmhouse by themselves. We renovated the large attic, put in a kitchen, and although we shared a common entrance, Teresa and I essentially had the second floor and attic while Mom and Dad continued to live on the first and split-level floors.

The whole arrangement was as illegal as sin, but nobody could spot the internal changes since nothing was visible from outside. The reason it was illegal was that we did not get any building permits and certainly such an arrangement would have required the expansion of the septic system. But we exercised our don't ask, don't tell option and lived comfortably and happily in our penthouse for seven years. Lovely years. Cheap living years. Wealth accumulation years.

When my grandmother, who lived in a small mobile home right outside our yard, passed away, Mom and Dad upgraded to a much larger, better built mobile home on that site and Teresa and I, with our now two children, took over the big farmhouse.

Because of this arrangement, Teresa and I were able to enter farming with no debt and live on literally a couple of hundred dollars a month. We live in an age when start-up businesses are supposed to indebt themselves for decades before the first sale or service item goes out the door. In other words, the old idea of starting the bootstrap business with no capital but keeping indebtedness and overheads low seems to be falling into disrepute. Now everything has to be with a big splash, which incurs a big debt.

If we had had to build a house, we would still not be farming full-time because we would both be working off-farm jobs trying to pay for the house. Our family did not have a big bank account. We did not come to this with independent wealth. The farm was very much a start-up business. The soil fertility would not let us produce more than about 15 calves a year. We weren't doing poultry or hogs at the time. The entire gross farm receipts in those first couple of years were under $15,000. Out of that we had to pay fuel, taxes, repair, insurance, and maintenance. It was tight.

But we did not do any outside entertainment—no eating out, no movies, no theme parks, no vacations. We grew virtually all of our own food. We grew all of our own home heating fuel. We drove a 20-year-old vehicle, and only filled the gas tank once or at the most, twice a month. We bought all of our clothes at the thrift store. Remember, without a front end loader, I hand shoveled 100 tons of compost every spring—that's five tractor trailer loads in case you didn't know. Just thought I'd remind you.

By living cheaply, we squeaked by until the fertility and sales began picking up. And today we are leveraging those frugal years with more capital-intensive projects that we can now cash flow. Economists love to talk about the time factor of money, and that is true. Another aspect is the exponential value later in life by not having saddled ourselves early with indebtedness.

In other words, $5,000 a year mortgage payments at that time would have kept us from farming full-time. We would not have been able to do what we did by being here full-time; we would have been relegated to weekend farmers. By being here full-time, we were able to do in five years what Dad and Mom were unable to do in twenty. That's the compounded value of not paying interest.

A realtor friend told me that young couples spend enough on automobiles in their first ten years of marriage to buy a starter home. He said if they would buy modest automobiles and save the rest, they could be in a home mortgage-free within ten years. Delayed gratification is still worth enjoying.

I'm belaboring this point a little bit because I'm trying to lay a foundation for the next turn of the wheel. Daniel and Sheri married but our daughter Rachel was still at home—living in the second floor of the farmhouse. Duplicating the farmhouse deal that Teresa and I had done was not an option this time. But by now we had abilities and resources that we had not enjoyed twenty years before.

Daniel, being homeschooled, had spent several stints as a teenager working with adult friends who did construction. Armed with that experience, he had expertise that I will never have. He could actually build a house. Trust me, nobody would want to live in the house that I built. The owner-built home was an option that Teresa and I just didn't have. Secondly, by this time we had our own bandsaw mill. And we had an excellent sawyer—*moi*. You know the old saying: the only difference between men and boys is the price of their toys? True in my case. I just love watching those logs turn into nice boards. And the smell of that fresh sawdust is intoxicating. Who in the world would want to snort cocaine when you can inhale fresh sawdust? Woodworkers can appreciate this, I'm sure.

Armed with those two abilities, we decided to build a honeymoon suite that could be either added to later or exchanged for the big house whenever children came along. Teresa and I have no desire to grow old in this big farmhouse. Mom still lives next door and we heat both structures with an outdoor wood furnace. It's a cozy arrangement and extremely efficient. We use a common septic system and common well.

How much room does a newly married couple need? The idea was to build this house out of existing savings that Daniel had, with us helping if necessary. The whole goal was to keep it debt-free and use as much of our own lumber and expertise as possible. Although we are better off financially than we were twenty years ago, the farm is still the only cash cow around here.

We don't have Swiss bank accounts and we don't have independent wealth. What we have is what we have. The point being that if Daniel and Sheri spent tons of money to build a Taj Mahal, the cost affected all of us. And the cost would hamper our ability to build ponds, upgrade the tractor, or whatever. This is a completely integrated operation. We aren't islands, although we have some autonomy.

We decided that 720 square feet—24 ft. X 30 ft. was big enough. We picked a site closer to the center of the farm so this development could extend hot power and potable water—we hooked into the existing farm well with a booster pump—farther into the heart of the farm. The big farmhouse/homestead setting is not centrally located.

We debated about getting a building permit or not. We knew that the real leverage came from the utility company. We weren't ready to do an off-grid house—we aren't savvy about that engineering and we needed to get this up quickly. The wedding was six months away. Daniel believed, based on other construction projects he had done around the farm, that he could complete this bungalow in time for them to get in it after the wedding.

We did not know of any situations where a family was evicted from a house they were living in because it was not built up to code. I'm sure those stories exist, but we knew a fellow nearby who foiled all attempts because he was not hooked up to public utilities. Since we were going to be hooked up, and the power company would not hook up without a building permit, we agreed to go the licensed route. It vexed my righteous soul, but this was Daniel's project and he wisely noted that he had enough things going on in his life with the farm and a wife that he didn't need the pressure of having to look over his shoulder every day to see if a bureaucrat was coming after him. The problem was a house is a hard thing to hide. After much soul searching, we agreed to try—at least this one time—to be legal.

Then came the introductory statement to this chapter: such a small house is illegal. Period. Doesn't matter if you're a hermit. Doesn't matter if you can't afford a larger one. Doesn't matter if this compels you to be a pauper for the rest of your life. You just can't build a smaller house.

"Why?"

"Because of resale value."

"We don't intend to sell. We aren't even platting off a separate lot. It can't be sold or borrowed against because it doesn't even own the land it's on. Besides, if we want to build a house that won't sell well, why is that the responsibility of the government? We'll just have a house that doesn't sell well. Shouldn't it be my right to build a white elephant if I want to?"

"Code says 900 square feet minimum."

That's it. End of discussion. When Daniel told me about this conversation, you can only imagine my outrage. I still haven't gotten over it. If I want to live in a yurt of Yak skins, what right does the government have to tell me that's not acceptable? People all over this world—and in our own country—live in squalor. Some are homeless. Some live in underworld sewers. Shouldn't they be glad we want to live in a house?

I can take you to domiciles in this county that are nothing more than mountain shacks. People live in them, make babies in them, grow old in them, laugh, cry, strum guitars in them, sing in them, cuddle grandchildren in them. My goodness, entire cultures live in crude thatched-roof houses or tents made out of skins. A recent article in the newspaper told about a local lady who was evicted from her apartment because she couldn't pay the rent. She lived in her car for six months. What do you mean a person can't build a house less than 900 square feet?

Do you think that lady living in her car would have been happy in a 200 square foot cabin? The important point here is that this is not just some isolated requirement that has no bearing on other things. This requirement makes a simple project much bigger, much more costly, and keeps us from being able to do our farm work efficiently. That in turn affects what we can produce, and the price of what we can produce, and directly affects our local food system.

Without any room for negotiation, Daniel added a second floor to get the space. We didn't want to expand the footprint. He located the house on the southern side of a hill in order to have a nice big crawl space underneath for storage—lawnmower and such—and a ground floor entrance from the upper side of the hill. The southern exposure allowed passive solar gain.

Next problem: to be up to code, the excavation required for permitted drainage would be more expensive than a basement. In other words, the house could not just be put on a foundation that blended in with the hillside. The whole area had to be excavated. I can't remember all the whys and wherefores, but I do know that the cost of a basement was less than all the earthmoving that would have to be done. Now we're at three levels: basement, first floor, and second floor.

Well, if we have a full basement, let's make an interior access to it so we don't have to go outside to get to it, even though it has a nice ground floor entrance because of the hill. Problem: a basement with interior access is habitable space, and all habitable space must be fully code compliant before an occupancy permit can be issued. That means it must be finished off with electrical outlets, wall and ceiling covered, insulated.

Okay, forget the interior access. We'll just get to it from outside. Meanwhile, I had been milling madly for a month and had all the lumber stickered and drying in the hoophouse after the laying hens came out to pasture. We closed the hoophouse doors and it made a wonderful kiln—hot in the daytime and cool at night. Daniel's idea was a modified timber frame interior skeleton with exterior stud walls. We milled poplar for studs and used oak for the floor joists and main timbers.

Next problem: your lumber is ungraded. That's okay, as long as everything is upsized one notch. If the specs call for a 2X6, we use a 2X8. If the specs call for a 2X8, we use a 2X10. Now folks, let me tell you something. If you go down to the lumber yard and buy a fir 2X6, it's light as a feather and actually a 1 1/2 X 5 1/2. Our boards were full-cut oak. Nobody can tell me that our full-cut oak timbers weren't as strong as a wimpy dressed fir piece.

But we had the forest so lumber was easy to acquire. I just cut extra big pieces and we used more logs. For the central timbers, where specs called for a 10X10 of multiple fir boards nailed together Daniel has solid oak 12X12s. You could park a tractor trailer in this house. I think it's so heavy it wouldn't even float.

Now we come to the second floor, where we envisioned a simple A-frame roofline to get the 900 square feet. Wrong again. A simple A-frame would not offer enough head space to qualify for enough habitable space. To qualify, more than half the space must be seven feet high. More complicated carpentry. That meant we had to go with a knee-wall to get the roof high enough to qualify for the 900 square feet.

When this project first started and he had not yet gotten the building permit, Daniel decided that to save money he would install a composting toilet. I had a speaking engagement in Ohio about that time so Teresa and I drove up and stopped at Lehman's. We found a perfect every-agency-certified honest-to-goodness four-person composting toilet and bought it for $1,000. Still a lot cheaper than a septic system—not to mention the environmental benefits. In our old house, the gray water—laundry, shower, kitchen sink—goes out onto the ground.

We have a bed of comfrey growing around the discharge area to convert the nutrient-rich water to high protein leaves, and in turn feed the leaves to rabbits and chickens. The leaves can also be crushed and used as a poultice to eradicate planter's warts. We figured if we dealt with the sewage, gray water would not be a problem. Nothing in that is really hazardous to anything.

Wrong: the health department does not recognize composting toilets as efficacious for eliminating the septic system. Gray water must still go through a septic system, no exceptions. Even with the composting toilet, Daniel had to install a full septic system. Had we known that, we would not have purchased the fancy composter.

When the health department official came out to determine if the soil could accept the septic effluent, we showed him where we wanted the leaching field. It was on the hillside near the house, nice and close and well away from any riparian area. He had his little handy dandy soil color guide and our soil didn't qualify. Too light. I asked him why.

"I've been doing this a long time, and they keep changing the color grades for acceptability. I don't know why, because every day I go to sites that are being updated or expanded or retrofitted, and the soil is just like what you have here, and it works fine. According to our instructions, they can't work. But they do, and have been for decades."

He couldn't find anything in the area that would qualify. To understand the situation, realize that if we had been asking for permission to expand an existing system that was in nonqualifying soil, the expansion would have been granted without question. But since this was a brand new installation, we couldn't get any concessions.

Years ago the health department didn't use the soil colors. Instead, the officials would use a little auger to dig a hole and fill it with water. If the water drained out in a certain period of time, then the site was deemed acceptable. Twenty years ago when a neighbor family moved in, I worked with them closely to get their site established. Their soil is far lighter than ours is, and it passed fine. One technique that people used to use was to make the appointment close to lunch. That way the septic police would dig the hole and pour in water, then drive back to town for lunch. While he was gone, the applicant would take a soup can and bail out the water. When the official returned the water was gone and he granted the permit. Worked like a charm. But those were the good old days.

After refusing to grant a permit anywhere nearby, the health department official asked where else we could go. The only other option was across our farm road and into a field near the creek. It was much farther away and of course would require digging up the road to lay the sewer pipe. Fortunately, it was still downhill from the house, so no pumping would be required.

The official went down there and sure enough, that tested fine. Now we were in the riparian zone, which I would think is not where we would want sewage effluent.

I pointed this out to him, and I will never forget his response: "Riparian areas are about the only places that will pass the soil color any more. That's why all the residential estate developments are going in along creeks. That's why all the residential developments are down in the fertile bottomlands, the best farmland of the county."

Folks, I'm not making this up. Isn't that insane? This is the government health department, protectors of public health. What a crock. Same department that said an unwashed egg was inedible. Don't even get me started on these folks. What has our culture done to itself when no infrastructure concessions are allowed for composting toilets and then the sewage can only be dumped next to the waterways? How ridiculous is that? But that's what's legal.

By now time was running out. The wedding was fast approaching and all the extra work these regulations required had mired the simple project in complexity. Not to mention time and money. Daniel finally bought a cheap used camper and parked it next to the house site. He figured they could live in that long enough to get the basement dried in enough to move in without an occupancy permit—illegally, of course.

That's exactly what they did. They lived in the uninhabitable basement while they finished the first floor, which included the kitchen. Then they moved into the first floor while they finished the second floor, where their bedroom was. After two years and an extra $50,000, the house was finished. It is a wonderful house. But at 2,160 square feet it's big enough for their family and we may never exchange houses. We don't know what will happen to the big old farmhouse where Teresa and I now live, except I know we don't want to get old here. It's just too big.

Bottom line: what could have been a half-year project financed by savings and cash flow, with the work fitted neatly around the farm chores, became a monumental two-year, borrowed-money, financially-strapping endeavor that hampered production and created unnecessary tension for a long time. All because our choice of lifestyle and domicile was illegal as decreed by people who think they are more concerned about our welfare than we are.

One final installation was required in order to get homeowner's insurance: a propane wall-mounted heater in the basement to keep the water tank and pipes from freezing. Because Daniel heated the house with a woodstove upstairs, the unheated basement could potentially freeze. With that installed, banisters on the stairways, and all the molding installed, he and Sheri received their occupancy permit two years after moving in. What a relief that now the government actually agreed that the house was habitable.

With this story as a backdrop, then, what about building codes? Are they really necessary? I had a customer here one day when we were in the throes of this project who announced, "I wouldn't want to walk into a building that wasn't inspected."

This was after he toured all our farm buildings, went into our house, and purchased food from our illegal on-farm sales building. None of these structures was inspected. Our house was built in about 1750, long before the current regulatory climate hit our culture. Our barns, sheds, and outbuildings do not require inspection because in our area, farm buildings are essentially exempt. I guess nobody cares if a building collapses on a cow. Our sales building was put up in a day by our church group as a community project. We milled the lumber and it has served us well for years and years. Obviously, he had enjoyed uninspected buildings all day, and didn't seem concerned that they would fall in on him. He trusted me, and that was enough.

The problem with codes like this is that they stifle innovation and complicate what otherwise could be elegantly simple projects. Just for fun, let me describe my dream house. And after reading this story about Daniel's house, which looks entirely conventional, just imagine the kind of gaff I would get on this dream house.

First, it would be earth sheltered—built into a hill with a glassed southern exposure. Gray water would exit to a small purification swamp using hydrologic plants like cattails. The toilet would flush into a sealed tank producing methane. Effluent from the methane digester would go into the plant purification system. The plants could be composted or fed to animals. The methane would go directly to the kitchen, where it would run the cooking stove and oven.

One of the biggest costs in a house is the roof. I would get rid of that cost by installing a simple ceiling covered with a hoop house. The hoop house plastic would be the real weather-tight roof for the entire house. On our current farm hoop houses, we use a tough nine mil UV-stabilized laminated plastic that will last for 15 years. Because the house was built into the side of a hill, both the upstairs greenhouse and the downstairs living quarters would have a ground floor access.

During the day, the second-story greenhouse would collect solar energy which could be pulled downstairs into the house with a couple of squirrel cage fans. At night, the thermal mass from downstairs would naturally warm the air and it would rise into the greenhouse. That way all winter we could grow fresh vegetables upstairs. We could even have a couple of chickens up there for fresh eggs and eating the kitchen scraps. Because the earth sheltered house would be built into a hill, we could put our freezer and a root cellar underground for energy savings and storage right by the kitchen. And the house would be cool in summer because of the earth shelter.

Can you imagine the building inspectors looking at a greenhouse roof? I can hear them now: "What? What happens when the wind blows it off? Or what if it gets a tear?"

My answer: "Who cares? It's so cheap we can replace it every year and it would still be cheaper than the roof you would require." See, I'm into the decomposing house. I think the Indians in this area were right on when they just bent over saplings and had houses that rotted in about 15 years. That's about how long it takes for your family size to change anyway, so you either upsize or downsize as necessary and you use real time resources to meet real time housing needs. Nothing more and nothing less. Oh boy, now I'm getting really weird.

But that's the point. The codes force everyone into the same mold. They stifle true innovation. The codes make sure that whatever is built tomorrow will look like what was built today. And so we continue building stick houses with materials, plumbing and energy requirements that are not in any way connected to the place where the house is built.

Imagine a housing development where more than 50 percent of the building materials, all the water, and 50 percent of the energy had to come from the development's landscape footprint. Edible landscaping would replace Chemlawn. Little earthworm bins would replace garbage disposals. Cisterns would adorn house edges and catch all the roof runoff. We could do lots of innovative things if it weren't for these building code straitjackets.

I know somebody is thinking about the poor quality workmanship that would occur. On the contrary. By and large the workmanship would be better because the contractors could not punt to the minimalistic codes. Just like slaughterhouse inspection, the bureaucrat takes all the pressure off the business.

As long as a bureaucrat signs off on the permit, the builder or the processing facility bears no responsibility, And the shysters can hide behind the bureaucrat's skirts saying, "Everything we did was up to government specifications."

Translation: "The paperwork was all filled out. It's not our fault."

I suggest that on the day the processing facility must stand on its own reputation for product that is being analyzed by private detection agencies, the quality of work will go up. And the same is true in construction. The day a roof collapses due to shoddy workmanship, that builder is out of business. Suddenly these guys will be working to their own reputation rather than to the code. Codes tend to move everything toward a minimalistic standard, thereby taking the responsibility off the private sector.

The private sector will only rise to its responsibility. When the largest, most institutionalized, least artisanal elements dominate the code-writing fraternity, a gradual diminution of quality is the inevitable result. When independent people put their independent reputations on the line with an independent product, they bear complete responsibility for their reputation. Then the workmanship goes up.

Again, certainly slipshod work will be done. But at least it will be isolated instances of incompetence rather than industry-wide shoddiness. I don't seem to remember reading much about collapsing buildings before the codes were established. This is another case in which the cure is worse than the disease.

And in the final analysis, it affects how a local farm brings on a second generation to maintain the farm and serve its local constituency. In our case, because the house project became exponentially more onerous than necessary, it was responsible for shutting down our sheep production, kept me from writing another book, and stifled several other projects that desperately needed to be done. Everything relates to the food system. Everything.

Before we leave building codes, perhaps one more story will help drive home how capricious and stifling these things can be. We're currently working with a couple in our county who are trying to build a small federal-inspected slaughterhouse. It would be the first one ever in our county.

They have five defunct poultry houses on the property. Their farming history is a storybook boom-bust illustration. They wanted to farm full-time. Poultry was moving into the area and money for building these CAFOs flowed freely from every lending institution, especially government-affiliated outfits. These monies were allocated by politicians bowing to save-the-farmer pressure, all the bleeding heart farmland preservationists.

A couple of years ago a large article in our local newspaper quoted a local farmer who praised the American Farmland Trust for keeping his CAFO poultry operation in business. In the big scheme of things, why is a CAFO more land friendly than a strip mall? I would certainly rather live next to a housing development than a CAFO, with its stench and fecal particulate exudates. For these farmland preservationists to make no distinction on the type of farms that should be saved is to refuse to recognize that some farms are helpful for the landscape and some are like open wounds.

But these nonprofit organizations are playing to heartstrings that don't want to learn very much about the problem. People ante up money in donations as guilt assuagement, never realizing that the nonprofit org skips lightly over the truth as easily as anyone. It takes effort to differentiate good farming from bad farming. The effort to differentiate farms worth preserving from those not worth preserving is too much for farmland preservation charities. It just complicates the fund raising drive. Easier to just plead: "Please help save farmland." Since limiting the preservation help to only good farmers is hard to do, it isn't done. And these organizations prey on shallow sentimentality, raking in the money to save farming. The result is that factory farming gets preserved, not community friendly farming.

These outfits could use their clout to make a great statement about the kind of farmland that should be preserved. The kind of farmers who should be preserved. But instead they collect and distribute these donations indiscriminately. In all honesty, I want factory farming to go out of business. When I hear about a struggling factory farmer, I want him to either quit or change. I don't want him bailed out to put another half a million chickens through his confinement house. What is noble about enabling another half a million birds to be disrespected, abused, and then fed to more nutrient-deprived people?

What these farmland preservationists do is disconnect farmland from farmers. What good is saving farmland if there are no farmers to farm it? Do we really want more wasteland? I know a fellow who signed up for one of these conservation easements only to be unable to pasture chickens a few years later because the portable field shelters were considered new farm construction. His preservation easement precluded any construction that would add a structural footprint to the farm.

Talk about confining. He couldn't built another house for a family member to come onto the farm. He couldn't add a pastured poultry operation. He could not add a greenhouse. I'm not a fan of these farmland preservation efforts, because the farmer then signs over to someone else the authority to determine what is farmland consistent or not. In my travels, I have found a virtual cornucopia of farm-friendly enterprises that are considered by most well-bred blue hair preservationist types to be incompatible with farming.

A farmer living on a main highway in Indiana added a restaurant. Now the farm is viable and employs many extended family members and community residents. Otherwise it would have been sold for development. Another family was facing bankruptcy and instead converted an old bank barn into a miniature golf course. Now the farm employs nearly a dozen family members and offers corporate recreational opportunities on a picturesque farm. And the farm caters its own food for the meal.

Farmland preservation doesn't work unless it encourages adaptability on the part of the farmer to perpetuate a viable business. I think about the farmland preservation effort like I do about the organic certification effort. If all the effort and money spent trying to preserve farmland had been devoted instead to breaking down the illegalities I've articulated in this book, we wouldn't have a farmland preservation problem. The question is not whether or not preservation efforts have saved any farmland, but rather what efforts would yield the most efficacious and long-lasting results?

Virtually all efforts accomplish some good. Remember, the road to hell is paved with good intentions. It is a question of bang for the buck, of yield per ounce of effort. And I suggest that all the preservation efforts created in this country would be dwarfed in yield when compared to the changes enabled by freeing up farmers to access their local communities with a vibrant cottage-industry food system. Seeing what I've seen, I have a hard time getting excited about any effort that proposes to solve things from the top down.

Back to my friends with the five defunct factory chicken houses. As young, aspiring farmers, they borrowed the money and built five houses. As they describe it, the first couple of years were quite good. Then the margins began to drop. And drop, and drop. Within a decade, they weren't making enough money to stay in business. They liquidated the houses and worked off the farm to finish paying the mortgages and get debt-free. They accomplished that.

Meanwhile, the empty houses sat. And sat. Then he found out how much his Angus steers were worth if he could put them in neat little packages. He began converting one of the chicken houses into a slaughterhouse. All went well until the building police showed up. A chicken house in our area needs a 25-pound-per-square-foot roof. This is the load bearing requirement.

The problem was that even though the county zoning police issued a special use permit for the retrofitted use, the building police categorize a slaughterhouse as commercial. The building code requires all commercial structures to carry a 30-pound-per-square foot roof load. The fact that these roofs had already carried 25 years of snow and were structurally sound did not matter. The fact that he was not attaching anything to the ceiling to add any load did not matter. What mattered was that a commercial building must be 30. Period. End of discussion.

Amazingly, if he slaughtered only his own animals, the county considered it a farm building and not commercial. If anyone brought their animals to him for processing, and he charged for that service, then he went under the commercial code. Do you see how these regulations have nothing to do with function or fact? Or safety? They are just arbitrary requirements. If the building codes deemed a 25-pound roof acceptable for the farmer, the inspector, and his workers to be in, why wasn't it okay for the same people to process their neighbor's cow? The building police would not budge.

Here is how he solved the problem: He will buy animals from customers for $1, process them, and then sell them back to the customer for $1. He will charge only for the processing service. That way he is processing only his owned animals and is not considered commercial. In this case, the federal inspectors couldn't care less about the load bearing capacity of the roof. All they care about is impervious walls, carcass hoist height, temperature in the chill room, and shiny stainless steel everywhere.

Knowing that his days may be numbered to continue this charade, he's hoping to increase the roof truss strength from retained profits as the business gets established. But what an absurd, unnecessary, emotionally, and financially draining requirement. It's obscene.

When the U.S. goes to war halfway around the world to extricate people from a tyrant, I wonder how many Americans understand that tyrants exist right here. They are fellow Ruritans, Rotarians, Kiwanians. They wear pin-striped suits and sit behind big desks collecting government paychecks. They tyrannize small businesses. They shut down local food systems so that millions more Americans believe their only alternative is to shop at big box stores.

This was the whole point of Patrick Henry's famous "Give me Liberty or Give me Death" speech. Spoken, incidentally, after watching an unlicensed preacher being flogged in the street by the Church of England dominated power elite. Tyranny comes in many forms, shapes and styles. I wonder how many Americans have been killed by policies that deny them alternative foods, medications, wellness practices?

This is not an academic question. It touches right where we live, how we live, and how we eat. The more difficult it is to access local food, the less free we are to acquire such food.

If, as every private and governmental report says, the vulnerabilities of our food system, both for pathogenicity and bioterrorism, are centralized production, centralized processing, and long distance transportation, then how many people have been or will be harmed by this system? The fact that decentralized production, decentralized processing, and short distance food transport are the antidote for these vulnerabilities suggests that policies and bureaucrats who enforce them to keep the antidote from seeing the light of day are de facto tyrants.

Anyone who thinks I am overstating the case has not had a visit from these people. They enter farms unannounced and rummage through private refrigerators, mail boxes, and take pictures indiscriminately. They gun down animals on the flimsiest of excuses. They seize phones, files, customer records, and product without offering compensation. In every sense of the word, they are tyrants.

And frankly, I have a hard time becoming concerned about alleged tyranny in other cultures when I see the kind of assaults against our own people, our own local food systems, and our own well being. I wish all the talk show hosts and pundits who think we should send U.S. troops around the world protecting the liberties of other peoples would understand the terror that farmers like me feel every day when some official arrives on our doorstep to tell us we are not in compliance. This is not paranoia. It is a fact of life when trying to create a free and righteous food system. It's domestic tyranny and terrorism, fully licensed and sanctioned by the U.S. government.

Just try selling raw milk or farm-slaughtered T-bone steaks to your neighbor, and see what happens. It's not pretty. To terrorize community-friendly and nutrient dense local food systems like the current regime does is to criminalize the answer to everything that is wrong with our industrialized food system. To outlaw healing is despicable beyond description. The passion with which food police confiscate, obfuscate, and terrorize is on par with the worst elements in any tyrannical regime the U.S. government labels as evil. Talk about the pot calling the kettle black.

Do I overstate the case? Have I gone off the deep end? Time will tell. When customers' children have life-threatening seizures and other reactions to toxin-laden government-sanctioned food and our food cures all those problems, who is the real bad guy here? Us? Or the government that says our food is illegal and these people, these desperate children, can't have it? The sooner we recognize the moral dimension, the holy dimension to a free food system, the sooner we will join the right side with verve and passion. Will you join me? I hope so.

Chapter 17

Insurance

"If you sell one processed chicken to an individual, we will not even cover your residence under your homeowner's insurance policy," said the insurance agent from Farm Bureau Insurance. This supposed friend of the farmer and insurer of rural America had been our comprehensive insurer for many years.

But someone found out that we were direct marketing, and that was the end of it. I remember those conversations well because they helped me understand that the American Farm Bureau Federation is the altar boy for the church of industrial agriculture. While this largest of all farm organizations touts itself as a friend of free markets, it is the first to impugn farmer direct marketers and ask for a host of regulatory bureaucracy to police such unconventional aspects of the food system. The Farm Bureau is a perfect example of limited free market. When its adherents say free market, what they really mean is heavily regulated markets with special concessions for large corporate interests.

Please understand the import of the insurance company's position. They were not saying that if we engaged in direct consumer sales that they would not cover us for product liability. They were saying that if we engaged in such activity, they would

234

not honor their commitment to cover our losses if our house burned down. That is the degree to which industrial agriculture fears anything outside its control. Product liability now dominates every market access and every discussion among direct-sale farmers.

About the same time, we were beginning to offer pay-per-day hunting. Hunting leases have been around for a long time. A group of people lease a piece of land for the season, maintain posted signs, and police it for the landowner. For this area, our 550 acre farm, was a sizable piece of land in one block and therefore an attractive hunting destination.

Over the years, public lands have gradually lost favor with hunters because they do not have the landscape diversity necessary to attract wildlife like farmland does. These public lands, both national forest and state game management areas, are essentially maintained as repositories of timber to keep lumber prices low. Notorious below-cost-timber sales suppress the price of trees, which makes them less valuable and therefore poorly stewarded by private landowners.

In Virginia, nearly 70 percent of all forestland is classified as NIPF (Non-Industrial Private Forests) but it only accounts for 30 percent of the harvest. The primary reasons for the low rate are the low value of timber and the poor stewardship of these forests. The low value means that landowners tend to put off selling anything because the income doesn't justify the hassle. That means that when these forests could benefit from thinning, it doesn't happen, which retards the growth of the good trees.

The second thing the low value creates is a disrespect for the resource. Rather than caring for their forests, farmers run their cows in them, which hurts the trees that are there and kills any regeneration. In our area, most farmers view forests as just something that gets in the way of the hay mower. On our farm, we judiciously harvest to maintain openings that sprout blueberries, blackberries, and succulent browse. This stimulates diversity and wildlife.

The dozen ponds we've built, plus fencing out riparian areas, and fencing out the woods, have all created prime wildlife attractions. We realized that earning some pre-timber income from this multi-decade investment would be another income stream and help keep the wildlife healthy through hunting.

About a decade ago we courted the crew chief of a tree chipper crew that trimmed brush and trees along roads and power lines to begin routinely dumping his loads of chips here at the farm. We bribed him with some cash that he just sticks in his pocket and learned that he was an avid rabbit hunter. One day he asked if he could come out with his son and some buddies to hunt rabbits.

Realizing that this was another pot sweetener to keep us on his carbon-dumping route—we are fiends for biomass for composting and bedding—we said yes. He showed up the next Saturday with about half a dozen rabbit dogs, three pickup trucks, about four men and that many young boys. They spent the day and had a wonderful time. They came back the next Saturday and repeated the fun. To cut to the chase, since the rabbits have been hunted hard, we have seen an explosion in the population.

I remember well as a youngster how seeing a rabbit during morning chores merited a spirited story at the family breakfast table. Today, we routinely see nearly a dozen every morning. And we have more fox and coyote pressure than existed in those days. The only thing I can attribute it to is the healthier population due to the hunting pressure. These fellows have continued to come and we've watched the boys grow into young men. They eat everything they shoot and the dogs are great entertainment. When they strike a fresh rabbit scent, their yelps fill the air with music.

For deer, which is the prime game animal in this area, we looked for a group that would lease the place for the season. We couldn't find one. Through discussions with a couple cosmopolitan hunters who did like to come out, we eventually hit on a new idea: a per diem rent. That way a person didn't have to make a big financial commitment for the season, and only paid to hunt when he hunted.

When people lease land, they feel obligated to use it a lot to get their money's worth, and that sometimes conflicts with other responsibilities—like work and wives. This was a new concept and about twenty hunters in the area went for it. We had a get-acquainted meeting and they each signed our protocol sheet. I created a detailed map of the whole place and designated about twenty sites scattered over the 550 acres. I built a check-in station so that when they arrived, they checked in on a numbered area, indicating where they would be hunting, and dropped the money in a lockbox.

We never had more than a dozen in a day, but it was a wonderfully flexible system. Some guys only came one or two days and others came a dozen times during the two-month hunting season. Nobody paid for more time than they used and we could easily police the whole operation through the sign-in process. Nobody ended up on top of anybody else because everyone knew where everyone else was. Farm Bureau Insurance was fine with this setup. The fact that we had a dozen hunters running around the woods shooting things didn't bother them at all. But that one plucked and gutted chicken was dangerous. So dangerous that they would not even cover our burning house due to the hazardous nature of someone buying a side of beef or a pound of sausage from us.

We checked around with other insurance companies and found a local outfit that had no problem with the direct marketing. But the thought of hunting customers paying to roam our woods sent them into apoplexy. If we didn't charge, but just let anyone come on the property who wanted to—drunks, irresponsible goons, unsafe gun handlers, fence cutters, trespassers, poachers—the insurance company was fine with that. If we let anyone come on the farm with guns a'blazing, and they shot someone or tripped and broke an arm, the insurance company said we were not liable as long as they were here for free.

But if we charged one red cent for the privilege, then we were liable and by extension, so was the insurance company. We were much better off, from an insurance standpoint, to have a drunken party of gun-slinging buffoons staggering about the place for free than a carefully screened, policed, responsible group who paid a penny to be here. Yes, they paid more than a penny. But the silliness of a proposition is most obvious around the extremes.

And the insurance company didn't care if it was $100 or a penny, a paying customer meant liability, and they wanted no part of it.

Let's get this picture right. We had two insurers. One didn't mind if we had a hundred hunters stacked on top of each other paying for the privilege, but were scared to death of selling one table-ready chicken or one pound of hamburger to a customer. Even if it was federally inspected.

The other company didn't care if we sold a ton of backyard-slaughtered or kitchen-produced food to people, but were scared to death of one hunter who paid a penny to roam our woods. Even if he carried his hunter-safety course diploma with him. These were our only options. And the crazy part about it was that our regular homeowner's policy was at stake. These issues could not be separated from our homeowner's policy. They wouldn't let us put a disclaimer rider of non-coverage on the insurance contract. It was all or nothing.

We realized the hunting potential didn't hold a candle to the direct marketing potential, so we went with the company that didn't like hunting but liked food customers. We went along fine for several years until our agent saw our farm featured in a national magazine. Trust me, our neighbors don't have a clue that we've been featured in national media outlets. A prophet is never loved in his own country.

"My underwriters are getting nervous about your exposure," he said. Not being an insurance afficionado, at first I wondered where I'd been spotted running naked. Exposure? I know that lots of times folks think earth muffin farmers like me get energized by running naked through the woods on moonlit nights to commune with the cosmos. But I couldn't remember having done that lately.

The insurance agent explained, "Exposure is your risk. We don't mind accepting some risk, but when you begin doing the volume of product you're doing, we get a little concerned. So I suggest we sit down and talk about a product liability rider on your homeowner's policy so both of us will have some protection."

Up until that time, I didn't want product liability protection because statistically nobody sues anybody who isn't covered.

Understand that the people doing the suing are lawyers. In other words, Joe Blow customer doesn't walk into a courthouse and file a suit. At least not very many. Most of them stop off at a lawyer's office first. Aren't you lawyers glad of that?

The lawyer, in my totally prejudiced opinion, is far more interested in the amount of recoverable assets than about the merits of the case. If it's a big pot of gold, the lawyer moves forward with the suit even if it's spurious. See, I know some big words too. Don't fool around with me, you high steppin' attorneys.

When the attorney finds out that the person being sued has no insurance, the whole system breaks down. Because now it's not attorneys talking to attorneys, it's attorneys talking to people. And that doesn't go as smoothly. If you study averages, the chances of being sued are in direct proportion to the amount of insurance coverage carried. In other words, the more product liability insurance I carry, the more chance I have of being sued. I figured the best protection was to just not carry any insurance.

I know some folks right now are cringing and are absolutely convinced that I'm a reckless, rebellious, egotistical fool. But I think of this a little like the car safety topic we debated one year on the intercollegiate debate circuit. That was my extracurricular activity in college—better than all the classes combined. The old big cars vs. small cars debate, which one is safer. As it turns out, the bigger the vehicle, the less likely you are to be killed in an accident. If I'm in an accident, I want to be in a big car. The survival rate is double compared to being in an accident in a small car. That's statistically proven.

But, my chances of being in an accident are half as likely if I'm in a small car because small cars are more maneuverable. I'm half as likely to ever be in a life-threatening collision if I'm in the small car. The fact is that while this debate rages, the two statistics cancel each other out. Go ahead and drive the big car knowing that if you get clobbered, you'll probably walk away from it. Go ahead and drive the tiny car and don't be scared because you can zig-zag your way out of harm's way.

See, these issues are not just always cut and dried. The fact is that risk is part of life. Every day I decide to get out of bed in the morning, I put myself at risk. In fact, even if I decide not to get out of bed in the morning, I put myself at risk. Who's going to pay the light bill, for instance? Who's going to eat Teresa's fried eggs and sausage? Life is risky; you can die from it.

Think of how much progress has been made because someone decided to take a risk. What if Jacques Cousteau had been afraid of risk? Would my generation have grown up with such a fascination toward the ocean and its cornucopia of life forms? I remember when my fellow employees at the newspaper learned I was leaving and coming home to the farm. A small farm. A young family. Every single person thought I was throwing away everything in reckless abandon. They all knew—especially the part-time farmers—that "there ain't no money in farmin'."

We can't make risk-free decisions. We can't live risk-free. Is it more risky to bet that I'll never be sued by refusing to join the insurance-litigation merry-go-round? Or is it more risky to buy insurance and hope that the person who will inevitably sue me—because such insurance attracts suits—will not have as good an attorney as my insurance company? Most of us never think these things through. We just go on willy-nilly with whatever our civic clubs friends say, and that perpetuates the status quo, and that isn't always the best thing to do.

If we don't occasionally break with standard practice, when will standard practice be brought into question? Or when will it cease to be standard practice? And if I don't break with it, who will? A stampede starts with one individual.

We were back to facing cancellation of our homeowner's policy if we didn't put the product liability rider on. We put the rider on—reluctantly. Little did we know that this issue would have surfaced shortly anyway, with farmers' markets.

When we wanted to join a couple of farmers' markets, we had to fill out paperwork proving that we had product liability insurance before we could participate in the market. This is becoming a much larger issue across the nation as more and more farmers' markets require coverage for their vendors.

As the consumer interest in alternative food grows, larger interests seek food products from local farmers. The problem is that these conventional food interests slap requirements on the initial transaction that often preclude the sale from ever occurring. Most retailers and distribution networks require at least $2 million in product liability coverage. Many small farmers can't even find coverage, let alone pay for that amount. It's important to understand that this is a requirement before the first sale, not something the farmer can grow into.

In other words, if we could have a scalable premium, based on sales volume, that would be doable. If I only sell $100 worth of product, for example, my premium is $1. But like too many regulatory overheads, this one is nonscalable. In other words, the initial $100 sale requires a $1,000 premium. The high overhead for the introductory sale is what makes this whole issue difficult. Not liability per se, but the inability to tailor the coverage to the sales volume.

By now, I hope anyone who has read to this point would intuitively understand that product liability insurance is much easier to get for government-licensed food than it is for local farmer-produced and processed foods. To put this in perspective, the insurance companies put their faith in a system that depends on bureaucrats to know the truth, codify the truth, and dispense the truth from inside the Beltway using a labyrinth of far-flung offices to police a stockholder-driven industrial food system. The insurance companies do not put their faith in the inherent accountability implicit in a customer purchasing directly from the farmer.

One system counts on mountains of paperwork, institutionally-trained experts, machines, cheap labor, bar codes, reactionary and spontaneous buying, and checkout lines. The other counts on transparency, relationship, contemplation, and high levels of personal touch. To say that one is risky and the other is not is ludicrous in the extreme. The truth is that they both carry a degree of risk.

I hope that by putting the difference in this context, everyone can begin to grasp the profound philosophical connotations of a cultural norm as basic as product liability insurance. As soon as we posit liability insurance as a necessary component of the food system, we buy into an entire conflagration of intrigue: wining and dining at the top levels of government; hot-shot attorneys dissecting legalese; inspector gumshoes receiving their marching orders; chief operating officers cranking out secret memos; industrial farms burying catastrophes and hiding pollution.

And now it's time to scream, "The Emperor has no clothes!" The system does not protect people from the things they fear most. It does not create secure warehouses and trucking lines and ports of entry. It does not reduce contamination by spreading out production into environmentally-sensible component sizes. It does not check for bacteria inside eggs. It does not monitor hand washing after bathroom use.

The system in which everyone places his faith is a charade. Oh, it fills out reports. It creates mountains of paperwork. I attended a food inspection seminar recently and the government inspector said, "All that matters is the paperwork." I visited a federally inspected facility and looked at their pre-start inspection charts. These are charts to fill out every morning to make sure the equipment is clean enough to use. These reports were filled out for the entire year. The office staff just duplicated a stack of them, with the same infractions noted to give the semblance that things had been checked, and they moved one sheet from the "To Do" file over to the "Done" file every day. Every day it's the same thing. Now aren't you relieved that the inspection requirements really work?

"Infractions Found." "Corrective Action Taken." "Policy to Prevent Repetition." The plant personnel are entirely on their own to determine if there was a breach of sanitation. The personnel are entirely on their own as to whether or not it was corrected. And the policy against repetition is simply that they will look at things when they come in again in the morning. All the bureaucrat does is check that all the squares have been filled in. If the paperwork is in order, that's all that matters.

Let's get it through our heads once and for all: you can't legislate integrity. A person either has integrity or not. You can take two people, read them the same protocol, and one can do a super job and the other creates a sloppy mess. That's human nature.

The things that the government agents go after are miniscule compared to what they let go with impunity. After 9/11 farmers could hardly get into Washington D.C. with delivery vans or trucks. Big tractor trailers with corporate logos on the sides were and are never pulled over by the police. But anyone with a plain truck was and is harassed unmercifully. One of the biggest untold stories about 9/11 is the number of farmers that went out of business around New York City due to prohibitions against delivery vehicles entering the city. Small farmers don't drive tractor trailers decorated with customized paint jobs.

We small farmers buy secondhand trucks and try to keep them together with baling wire and duct tape. In the name of preventing terrorists, these poor-boy trucks were and are targeted by the authorities to the point that farmers finally just give up trying to get into the city. And yet small farmers are the least targeted by bioterrorism.

And it doesn't end there. One of my favorite stories came out of Illinois. A Kettle Korn vendor cooked his popcorn at home and transported it to the local farmers' market in his mini-van. The cooker and related hardware are just a headache to transport. Since it was such a short drive, he found it much more efficient to pop the corn at home and shuttle it to the farmers' market stand in his minivan. But the local health department official stopped him because "a bioterrorist might taint the popcorn in transit. You will now have to pop it on sight."

And yet how many lonely nights do prepackaged microwavable dinners spend in tractor trailers traversing interstate highways in the nation's hinter lands? Nobody checks them. How about all the trucks idling for a few hours at rest stops and truck stops while drivers sleep to remain legally awake? Farmers can go out and indiscriminately spray pesticides, herbicides, shoot their cows up with systemic parasiticides. We can irradiate food so that at least the poop we eat will be sterile, and all this without any label.

We can import food from countries spraying and feeding things that have long been banned in the U.S., and no label is required to indicate that it came from a foreign country. We can feed steroid ionophore-laced Rumensin and Bovatec to cows without any licensing and no inspection. Farmers in my area today, in early 2007, as I write this, are still feeding chicken manure to cows, completely legally.

Folks, the things we're moving heaven and hell to stop are not the things we need to be concerned about. The stuff that we turn a blind eye to and don't even care about is the bigger problem. We really do worry about the wrong things. And we devote precious resources inspecting, measuring, and vilifying the antidote to the wrong things.

Decentralized, community imbedded processing and distribution are absolutely the answer to the bioterrorist threat. But we put them out of business while we further populate slaughterhouse cities with increasing thousands of non-Americans. Don't misunderstand. I am not against immigration. But if we're going to have laws, we should apply them—fairly. I guarantee you that if I hired an illegal alien, I would be in jail tomorrow. But not the top brass of large corporations. That's what I mean. Our culture has lost its rule of law and become a respecter of persons, and that's a shame.

A republic that becomes a terror to righteousness and operates on the basis that might makes right is not better than a dictatorship that operates on the same principle. A democracy that worships money and power is no better than a socialist society that holds the same values. To run roughshod over the small people, the odd people, the opt-outers is the ultimate mark of tyranny. I'm afraid our republic is perilously close to that brink. Things are wacko.

Now for a breath of sanity. I propose a national Food Security Act that would have two parts. The first part would offer guaranteed freedom of choice to every American citizen to decide what to eat, and legalize every source. In other words, people would sign a "I Am Responsible for My Food" waiver that would give them the right to opt out of government-sanctioned food. And anything such citizens wanted would be legal to sell on the part of the farmer.

What I'm after here is a parallel food system. Essentially, it is modeled after home schooling laws, which allow for a parallel educational system. Our nation is richer for having enabled home schooling to survive. I suggest that we would be richer, too, had we allowed Indians to survive. We will be richer if we allow indigenous food systems to survive.

The problem is that the government feels responsible for every consumer's decision. Somehow we need to let people formally absolve society of the responsibility for their decisions—only the people who sign the freedom form. Creating freedom for autonomous food decisions is as American as anything I can imagine.

What good is the freedom to worship, the right to keep and bear arms, and freedom of the press if we don't have the freedom to choose what to feed our bodies so we can go sing, shoot, and speak? The only reason the founding fathers did not grant the freedom to choose our food was because it was such a basic, fundamental personal right that they could not conceive that special protection would be needed. Granting citizens the right to choose their food would have been similar to granting them the right to see the sun rise, or to breathe.

The fact that we even have to create an apologetic to defend the right to choose our food in this country shows just how far toward tyranny we have moved. Of course, I'm sure the founding fathers could not have conceived the bloated bureaucracy and the mountains of regulations that currently emanate from the government.

One of our former apprentices called me recently and said that in his state of Iowa, only eggs washed in chlorine may be sold to restaurants. Chefs, charged with the well-being of their patrons, cannot even acquire, legally, a non-chlorinated egg. Be assured that the next step will be to mandate chlorine-only eggs to individuals.

Somehow, folks who want to choose what to eat must be able to exercise this most fundamental, visceral, instinctive right. I don't know how these bureaucrats can attend Mass and Communion, how the officials at the American Farm Bureau Federation, and the industry trade organizations can sleep at night knowing they have denied Americans the decision-making autonomy over their own meals. And we Americans accuse other countries of violating women's rights or not being democracies. Unbelievable.

The reason these officials can live with themselves is because they really believe they are saving thousands of lives by denying people food choice. These powerful people actually believe, in their heart of hearts, deep down in the soul, that if you buy my beef, or my chicken, or my eggs, or the pound cake from our farm house kitchen, without first having it sanctioned by a government agency, you are imperiling your life. You are playing Russian Roulette. You are going to become a casualty, a burden to society, a nonproductive member of the proletariat, a liability to the culture.

Therefore, they see it as their civic duty, their ultimate Good Samaritan sacrificial deed, the most loving thing they could do, the quintessential Good Neighbor Policy, to save you from yourself. They really believe I as an unregulated local farmer/food producer, want to kill you. Or at least that my incompetence will kill you inadvertently. They believe they must protect you from me. And that gives them nobility as they go about their dastardly deeds. This is why the starting point on food freedom of choice is to strike at the foundation, the philosophical justification, that the government is responsible for your safety.

At least one person in society should have the freedom to choose her own food, even if its for a negative example. The waiver, then, would recognize that this is a freedom worth protecting. The few who wanted to exercise it could do so, and absolve all these noble public servants and industrial watchdogs their sacred protection ministry. And the food police could still sleep at night knowing that people voluntarily exercised their liberty to step out from under the government's umbrella of protection.

The second part of this legislation would grant a "Waiver of Liability" in that food transaction on all parties. The consumer who signed the opt-out freedom form would agree to not sue their food source for any reason for anything. That absolves the cottage industry pot pie maker, the farmer, the neighbor girl who made the pound cake, all from liability. The point being that when the consumer accepts full responsibility for her decision to purchase food that the government hasn't sanctioned, she also agrees to accept full liability for that decision. That allows the food police to sleep at night.

If she buys from an unsafe farmer, or patronizes a dirty kitchen, she waives the right to sue anyone for anything. No responsible food system can exist if the consumer holds all the marbles and the farmer-producer-processor is still completely liable. This puts the onus for checking things out squarely on the shoulders of the consumer. I know our customers would sign such a document in a minute because they understand that integrity cannot be legislated and they would love to see our prices drop, along with more variety offered.

The beauty of this plan is that it necessarily puts the opposition on the moral low road. Those who would oppose such a freedom have to argue that consumers are too stupid to take this responsibility. Their objective to deny freedom of choice could no longer be kept secret. And all their objections about farmers being sued, government's responsibility to protect the general welfare, and all the rest would be an obvious smoke screen.

Anyone who isn't willing to sign the Indigenous Food Freedom form can go to government-sanctioned food stores. And anyone unwilling to absolve their sources of liability can't purchase from Food Freedom sources. This enables an entirely parallel food system to exist. This way, those of us who want to slaughter our hogs in the clean grass of the front pasture can do so. If we want to cook chicken backs and necks into wonderful broth in a stockpot on a wood fire in the outdoor fire pit, have at it.

The industrialists cannot in good faith complain about losing market share, because if they do it will be an admission on their part that multitudes of American citizens don't trust the current government system. On the other hand, our side can say that if nobody takes advantage of this freedom, obviously no one wanted it in the first place.

The risk of sickness could then be assessed on its merits. People like me view ourselves as the greater protectionists, because we believe a more protected food system is one that encourages locally grown, processed, and distributed food. This turns the idea of who really is for protection on its head. Most people assume that those who deny freedom of food choice are the protectionists. I can assure you that the government food police view themselves as the ultimate food protectionists. In reality, though, they are the biggest anti-protectionists because they push the food system toward risky behavior: centralized production, processing, and long distance distribution. Shipping American whole chicken to China for further processing and then re-importing it is absolutely more risky than getting chicken from a neighbor farmer.

As the two parallel systems developed their track records, then, a valid risk comparison could actually be made. Currently, the paranoia surrounding an unregulated community food system is all conjecture. It's all about what could happen, might happen, or potentially happen because an unregulated system currently does not exist. All arguments against it are purely speculative.

It's like the liberals saying that the free market is responsible for corporate abuse and excess. Not true. We haven't had a free market for a very long time. I can't think of anything in the American economy that enjoys a free market. Perhaps eBay is as close as anything right now, and who knows how long that will exist? More conventional retailers are assaulting that venue with a vengeance. The tax-and-spend crowd hates eBay. True free markets simply don't exist for long because they generally begin displacing the entrenched system that enjoys regulatory oversight—a euphemism for political protection.

Even the rise of the petroleum industry was enabled by Prohibition. The religious right who created the Women's Temperance Union and eventually outlawed alcohol for a decade destroyed the imbedded, indigenous, farm-based, decentralized fuel system. The Model T Ford had a switch on the dashboard for gasoline or alcohol because many farmers and communities had their own alcohol fuel system. But Prohibition put the government in charge of alcohol and the historic, imbedded fuel generation infrastructure quickly vanished under criminalization. A tragic loss. Back to food.

The point is that we can't know whether unregulated community food systems have negative consequences until we try. We know what the track record is of a regulated, centralized food system, and it's dismal. Not just in direct illness, but in general food nutritional quality, taste, and texture. Not to mention pollution and rural economic and social devastation. Isn't it about time to allow an alternative just to see what would happen? If a bunch of people begin getting sick, like the current entrenched power brokers prophesy, then it will die a quick death.

But if its adherents are healthier physically, emotionally, and socially, then it will grow on its own merits. No contrived statistics. No special political favors. No grants, no corporate welfare, no free infrastructure. Just plain old merit. Isn't it at least worth a try?

What this would do is create a vibrant alternative to government food just like home schooling has created a vibrant alternative to government schools. And I'm not saying home schooling is for everyone. I'm not even saying that government schools can't teach anything. All I'm suggesting is that for those who want to exercise their autonomy, to exercise their op-out freedom, some way should be made for that to happen. And when an alternative, parallel system is allowed to exist, our culture is richer as a result.

And beyond that, I would suggest that preserving such a community-friendly food system would create more security, not less. Indeed, it would revitalize rural economies, returning on-farm and neighborhood infrastructure to the grassroots. The repository of power and money would shift, at least some, from metropolitan centers to rural areas, preserving the character of American hinterlands. That would be a good thing.

Chapter 18

Taxes

Forget the farm tax software templates. Our farm doesn't fit into anything. But try explaining these things to the accountant and he'll say, "Look, I don't want to know. Give me the figures and shut up."

For example, ponds are supposed to be considered capital improvements, which cannot just be expensed away. They should be expensed over several years. But we view ponds as a fertilizer expense.

Here's the rationale. Ponds grow more pounds of plants and animals per acre-foot than an acre-foot of soil. An acre-foot is something an acre in size (about 5,000 square yards—roughly the size of a football field) and one foot deep. One of the reasons water grows more pounds of living things than soil is because whatever grows in it doesn't have to expend calories defying gravity.

About 15 percent of the energy expended by plants and animals is to stand upright. The buoyancy of water virtually eliminates that energy requirement. Imagine the difference between picking up and pushing a 1,000 pound cow around compared to pushing a 1,000-pound boat floating at a dock mooring. As the pond grows vegetation and critters that

250

eventually go through their life cycle and die, their carcasses fall to the bottom. A nutrient-dense muck slowly builds that is better than any fertilizer.

On our farm, when we have a drought, we clean this muck out with our tractor front-end loader and spread it on the fields. I believe that a farm with enough ponds could be completely fertilizer independent by draining and cleaning one per year, say on a 20-year cycle. A typical farm producing our volume of food would normally have a fertilizer budget three times the size of ours. But instead of spending money on fertilizer every year, we build ponds. Why shouldn't our ponds qualify as fertilizer expense?

Looked at another way, the ponds provide irrigation. In our area, irrigation is virtually unheard of. In the 1960s many people in our community irrigated, but since then the droughts have not been as severe, until recently, but now nobody can afford to get back into it.

At any rate, the best way to fertilize the soil is with the sloughed off root hairs that the grass plant jettisons after being mowed, whether mechanically or by a grazing animal. Forage is always trying to maintain bilateral symmetry between what you can see and what you can't see. In other words, if we view the soil surface as a horizon, the plant maintains biomass balance on each side of that horizon. When the plant loses material above the ground, it prunes off enough root hair mass to create equilibrium. This jettisoned material adds organic matter, which is the key to soil decomposition. And that runs everything else. We call this pasture pulsing, this routine injection of jettisoned root biomass. It's far superior to sequestering carbon than forest systems.

In the dry summer, no matter how fertile the soil, once it gets dry it turns dormant. Without water, biological activity ceases. Among other things, grass is a perfectly designed solar collector. At no time in the year is more solar energy available for collection than in a summer cloudless sky—the kind of weather that accompanies a drought. Applying water can keep the solar collector working, which produces additional decomposable biomass, feeding the earthworms and soil food web. Again, fertilizer. We aren't buying it; we are building it.

Finally, ponds run our water system. If you buy a pump for your well, that's a legitimate expense. But we go up in the mountain and build ponds that gravity feed. No pumps. No pressure tanks. No electricity. Nothing to wear out or break. It's not fair that we can't expense anything when our system eliminates all the components that are properly expensed. In other words, if we would just buy the things that everyone else does, we could expense them. But since we substitute other items that are viewed as capital expenses, even though the total cost is the same, we can't legally write them off.

Beyond that, if we install a floating garden out in the pond, bugs trying to get to the vegetables land on the water and fish eat them. What a wonderful insect control. If we were buying insecticides to treat these bugs, we could expense that. Why then, if we build a pond to eliminate the bugs, can we not expense the pond as insecticide?

An unconventional farm selling directly to local folks has a host of these anomalies because a farm like this is not an industrial farm. And anything atypical drives the software people and the accountants crazy. As a farmer, then, I have three options:

1. Follow the book and pay a bunch of extra taxes.

2. Be honest with my accountant, pay him for his advice, and lose on virtually every argument—by the time I'm done paying his listening bill, I'd be better off just paying the extra taxes.

3. Keep quiet, call the invoice something that fits in our thinking, that's reasonable, and hope if and when we ever get audited that the arbitrator or judge is a local foodie.

I don't really like any of those choices, and truth be told we practice a hodge-podge of the three. Obviously we want to expense as much as possible. And it's not just short term expense vs. long term depreciation.

For example, what is the Eggmobile in our accounting? For you uninitiated, the Eggmobile is a portable henhouse that we pull around behind the cow herd. The chickens scratch through the cow manure, incorporating it into the soil, spreading it around, and picking out the fly larvae in the process. They also eat grasshoppers, crickets, and worms.

It's certainly a farm machine. But it's also our grubicide and sanitizer for the cows. It's a manure spreader. It's an insecticide. And it generates fertilizer from the chicken droppings. Besides that, it's also a free-standing profit center, a separate financial holon. Where the normal farm would spend money for grubicides and insecticides tabulated on the accounting sheets, we have zeros there. Which leads these folks to ask if this is a bona fide farm or not. It's all quite confusing.

How about vehicles? The farm owns our vehicles. But what about trips to church or to attend a local play? When you're selling locally, every trip is a potential sales trip. We have customers at virtually every venue we attend; and if we don't yet, this trip might be the one we pick up that other customer. Sitting in the waiting room at the eye doctor we often talk about the farm and may hand out a business card. Every trip and every activity is a marketing expense. We are never unplugged from the farm ministry. In this model, then, every trip is a business trip. Even a trip to the park will often include a conversation with someone about food, and eventually Polyface food.

I think we should deduct all of our living expenses. Here's why. The farm would not function if we didn't live here. If this were a corporation hiring a 24-hour security force, all the expenses associated with that would be deductible. As a side note, I always thought it was unfair that a business could send a bus to pick up workers and deduct it as a business expense. But if those same workers drove to the office, their mileage wasn't deductible. This is part of the discriminatory nature of the current tax system—it always favors the large corporate interests over the small ones.

If we moved to town and hired someone to stay in our farm house to provide security, we could deduct that expense—including the domicile. But since we live here, we can't deduct it. Those are the kinds of tax technicalities that large corporations find and utilize. But for a local food producer who has an integrated business/home/life, the tax code does not allow similar deductions.

Perhaps the most discriminatory tax is the property tax. This affects all farmers and is not unique to the local food producer, but I wouldn't be true to myself if I didn't broach the topic. Numerous highly credible tax studies have been done to show the tax bite in different sectors compared to the governmental services provided. Most of these compare residential and agricultural paybacks. In each case, the agricultural pays in about $1.30 in taxes for every dollar in services used. Residential pays about 80 cents for every dollar in services used.

The reason is that cows and corn do not need teachers or policemen. I've seldom seen a chicken in jail. People demand government services: plants or animals do not. Even with the land use taxation concession, agriculture still pays an inordinate share of the property tax burden. What this means is that farmland subsidizes residential governance expenses.

No matter how you slice it, that discriminates against farmland. All these programs to save farmland, from easements to trusts to purchasing development rights—why can't we get momentum behind changing the fundamental discrepancy in the property tax? Our politicians want to do everything but attack the most elementary financial unfairness against the farmer.

Farmland requires precious little in government services. Farmers install their own water systems. We install our own sewer systems. By and large our kids don't roam the streets at night because they are tired from doing chores. We don't get a thin dime to defray any improvements. If we build a shed, we pay for it ourselves. If we build a lane to get from one field to another, we pay for it ourselves. Perhaps I should put a clarification caveat in at this point: I'm talking about local property taxes, not federal USDA subsidies. For the record, they are absurd too, but that's another discussion.

But if a huge industry comes in, the county board of supervisors floats big tax-free industrial development bonds to help out financially. A 40-acre distribution center moved into our county a few years ago and it cost less for them to hook up to county water than it does for a county residence. That's corporate welfare.

Last year our county spent $500,000 on a feasibility study to build a mega-site for a huge industry—rumor had it as a Toyota plant—on farmland, including taking the farmland by eminent domain. But in this county a person still can't buy a T-bone steak that was born, raised, and slaughtered in this county. First it has to be exported to an adjoining county where a federally inspected slaughterhouse is located, and then re-imported to be sold.

If only half of the beef consumed in this county were processed here, it would generate some $30 million in local business revenue. That is the kind of true rural economic development we need. But politicians and their minion bureaucrats seem forever enamored of concrete, steel, and big everything.

Our little local custom slaughterhouse purchased a new walk-in freezer. Adding it to their existing building made it 12 inches too close to the public road—a violation of set-back requirements. Here is a multi-generational family business that has been a faithful taxpaying entity and service provider. This family pays their bills and is in every way the kind of self-starting entrepreneurial business that builds thriving local economies. But the county would not give them a variance. Nobody could even see the 12 inches with the naked eye.

But if that business had been some industrial yahoo outfit offering a wine and cheese hospitality night for the supervisors and planning department, the county would build them a road, hire the spouses of management to be teachers in the schools, forgive five years' worth of taxes, and take a farmer's land to make sure they had enough room to pour concrete and plant rebar.

I'm not against industry. I'm not even against big industry. And I'm not against residential development. But these things should carry their fair share. As it is, they enjoy a direct farm subsidy for the additional burden on government services that they require. Schools, jails, police, and rescue personnel all increase when more people come to town. And these are expensive government services. If farmland were not assessed at a discriminatory rate, perhaps more of it could be kept around to serve the community with wholesome local food.

But rather than address this fundamental inconsistency, greenies ask farmers to give up their autonomy through signing conservation easements. Oh, the platitudes about "protecting farmland" by folks who have no clue what reducing the tax burden to a fair rate would do to accomplish that end. Unfortunately, most of the people who understand this fundamental discrepancy – the greenies – are of the political persuasion that taxes should never be lowered. That even now we aren't paying our fair share. I never cease to be amazed at how easy it is to dance all around the problem, create additional government programs to deal with the problem, develop a new government agency to address the problem, rather than to just lay an ax to the base of the problem and cut it down. If one economic sector is shouldering an inordinate tax burden, fix it. Period.

How high should taxes be? I go to Genesis and the story of Joseph for the answer. Remember that Pharaoh had a dream in which seven thin cows ate seven fat cows. Joseph interpreted the dream as seven years of plenty followed by seven years of drought. Pharaoh was so impressed with Joseph's wisdom that he gave him the royal ring and made him second in command over all Egypt. Immediately Joseph built huge grain storehouses and began buying grain—a government grain inventory program.

Sure enough, the drought came, and with it famine. First, the people came and bought grain. Then they ran out of money so Joseph (the government) took their cattle in exchange. When they ran out of livestock, they deeded over their land. And finally they gave themselves as servants. By the time the famine was over, the government owned the livestock, the land, and the people.

And when Pharaoh asked what a fair tax would be, Joseph said the people should keep 90 percent of everything they produced and Pharaoh could have 10 percent. And that is what they did. Immediately, Egypt's grandeur returned and it was a prosperous civilization again. I suggest that if only 10 percent is adequate when the government owns the land, the livestock, and the people, then certainly that ought to be enough in our culture where the government doesn't own all those things. What is our total tax burden now? About 50 percent if all the taxes are totaled? That's obscene. And the most unfair portion is squarely on the most defenseless constituency in the country: honest-to-goodness farmers.

The reason the property tax is such a critical element is because most of a farmer's assets are tied up in real estate. For the first time in our civilization, real estate values have no relationship to the productive capacity of the land. Before the year 2000, fertile land sold for more than poor land. Not anymore. The viewscape is more valuable than the productive capacity.

These new dynamics are completely beyond the farmer's control. The inflated land value doesn't mean it suddenly will grow more grass or more zucchini squash. And unless the farmer decides to sell, the increased value is meaningless. It can't be tapped. It can't be turned into more hamburger or sausage. The additional paper equity is exactly that—paper equity. Unless and until the farmer sells, land value is meaningless.

To tie property taxes to land value, therefore, means the farmer is being unduly burdened by an arbitrary value that cannot be recouped. How would the average person feel if suddenly their car quadrupled in value and the personal property tax went from $200 a year to $800 a year? The car can't go any faster. It won't go any longer between oil changes. It won't carry more people. It doesn't get any better mileage. The increased cost cannot be compensated by the output of the car—unless the car is sold in the newly inflated market. But every other car is equally inflated. You can't trade up or out.

That's the situation facing farmers, and it especially affects farmers closest to residential development where a local food system is most possible. An all-out cultural effort to reduce farmland taxes should be the top priority for every greenie. Yes, that means reducing taxes. It is only fair.

Finally, the most despicable tax, in my opinion, is mandatory Social Security—or Social Insecurity. This confiscatory tax that returns pennies on the dollar compared to private investment accounts should be phased out.

This bite is huge, another program that started sincerely enough but became an entitlement completely out of control by the early 1970s. I have tried and tried to figure out how to get out of it without going to jail, and I haven't figured it out.

The Amish and Mennonites are exempted because they won that concession when our civilization wasn't so socialistic. If those groups today tried to get a waiver, I guarantee our political climate would not grant it. Why is it the government's responsibility to make sure I have money when I'm old? I know, I know. We're back to the "you won't do what's good for you so we have to do it for you" mentality.

Yes, I've read about elderly folks a hundred years ago that struggled to make ends meet. But they didn't starve. When taxes were lower, philanthropy and personal familial responsibility could be more freely exercised. With today's confiscatory tax structure, so much is bled off from the front end that the system has precious little discretionary private monies left. The system is completely broken and I would opt for throwing it out.

At least it should be privatized so the money could go into the real economy and not siphoned off to the government's general fund to be spent foolishly on building empires and buying $50 screwdrivers. When I think of how much money we've paid into this system and how little we'll get in return, it's immoral.

If the Amish and Mennonites can opt out, why can't I? I'll be glad to sign a contract that I will never ask for one red cent. If I starve to death in a gutter somewhere, that's better than saddling my grandchildren with a mandatory confiscatory "savings plan" that doesn't even return as much as a passbook savings account. Of course, if our civilization hadn't killed 50 million babies that would have been paying into the system right now the shortfall would not be as acute.

That is why we allow continued inflows of illegal aliens. We can't afford to stop the flow because these are the folks propping up Social Security. And they are doing the work that aborted babies would have been doing right now. We've executed our work force and must now accept whatever we can get. It's a sad state of affairs.

And now the death tax. Inheritance taxes fall heaviest on farmers. When Mom and Dad bought this farm in 1961, they paid $49,000 for the big farmhouse, equipment shed, pole barn, tractor, rake, baler, hay wagon, and 550 acres. Can you imagine? Today it's assessed at roughly $1 million.

I challenge anyone reading this to explain to me how my life is any different due to that $950,000 value increase. It doesn't grow one more pound of beef. It doesn't receive one more drop of rain. The house doesn't house one more person, or generate one more dime. The barn doesn't hold one more bale of hay. The tractor doesn't generate one more horsepower. The paper notion that this farm is worth nearly 20 times its 1961 value has zero income-producing merit. I don't give a hoot if the appraiser says it's worth $3 million. Big woop.

If it were assessed at $20 million, what's that to me? It has zero bearing on the productive capacity of the farm. It has zero bearing on the profitability of the farm. Ultimately, it has zero bearing on my worth—unless I sell it. Then I'm liable for capital gains. And we could talk about that, but that's a different issue. Right now, I'm trying to help everyone understand that the inflationary spiral on farmland values does not make farmers wealthy or more profitable. All it does is confiscate farms wholly or in part. If I elect to stay on the farm and grub out a living, why should society expect me to pay thousands and sometimes hundreds of thousands of dollars to the government just to earn the right to stay?

And even if I figure out how to escape these confiscatory death taxes, the meetings with attorneys and time required to figure it all out take a huge toll on the farm. Wouldn't you rather I be figuring out how to grow a better cow than spend my days stewing about how to dodge inheritance taxes? I actually think all inheritance taxes are wrong, but in this discussion I hope I can at least make an ironclad case for a farmland inheritance tax exemption. You could argue that apartment complexes that escalate in price are more rentable. In other words, the higher assessment has a relationship to the higher rent charge.

But on farmland, the same relationship does not exist. The inheritance tax promotes the notion that the day Mom dies, I should pay the government $250,000 in order to keep the farm. Where is that money supposed to come from? And further, by what rationale could any society justify demanding such a sum? If our culture has promulgated one great evil upon itself, certainly subjecting farmland to inheritance taxes is the most glaring example.

For the record, I am a proponent of the Fair Tax, as defined by Congressman John Linder and the libertarian talk show host, Neal Boortz. I agree that it is high time to abolish the regressive income tax. The whole idea of mandatory wealth redistribution is repugnant to anyone who understands rewards and punishments. Anything that punishes achievement and rewards laziness will destroy itself eventually. And the regressive income tax is doing exactly that to this country.

An across-the-board consumption tax is far more equitable. For me, the fact that the current tax code increases the cost of every consumer item in this country by 22 cents on the dollar is enough reason to abolish it. Many Americans don't realize that when this country was established, the federal government was to be financed by excise taxes, or tariffs. No income tax. Today, not only have we virtually abolished tariffs, but we have instituted a confiscatory domestic income tax. It's a double whammy against Americans, and is just one of many items destroying this nation.

While we're on the tax topic, let me throw one more out there. I think we should be able to earmark our taxes to certain governmental departments. That way the people directly vote, with their tax dollars, for the agencies they think best represent the good use of government. If you're a dove and despise the military, then earmark your taxes to the Environmental Protection Agency. If you're a homeless advocate, earmark your taxes to the Department of Housing and Urban Development. If you're concerned about foreign hordes overrunning our shores, earmark your checks to the Immigration and Naturalization Service.

And if you really think all those agriculture subsidies are important for food security, then earmark your tax check to the USDA. This way the American people could directly steer government spending. Government programs that failed to excite people would just die on the vine. The areas that people felt the most passion about could expand. It would be a true market-driven government. Let the Liberals and Conservatives duke it out in their earmarked taxes instead of smoke filled rooms where power players usurp the people.

Maybe we'd better get out of this chapter before I really tell you what I think. Here's the conclusion: all taxes are too high. But the tax code and tax burden fall inordinately heavy on farmers selling food in their localities. And the code acts as a straightjacket against innovation and fair payment strategies toward those of us who are square pegs in round holes. Those of us who actually believe selling to our neighbors is a good thing.

The Future

Chapter 19

Avian Influenza

"We're federal veterinarians from the USDA and we're here to take blood samples of your poultry," said the man in the sedan with blue government plates. The car, sitting on the grass in our front yard, had come to a stop between two picnic tables where our Polyface Intensive Discovery Seminar participants were enjoying dinner.

Unannounced, 6 p.m. on a Friday evening, the Avian Influenza Task Force veterinarians drove right into the midst of our dinner-on-the-grounds and demanded to see the chickens. I didn't know they were there because I was out in the back yard talking to some of the folks who had not yet been through the serving line. One of the attendees came around the corner of the house and said, "You might want to come around here. Two guys from the federal government are here to see the chickens."

I had about 50 yards and 2 minutes to collect my thoughts as I approached the car.

They had judiciously not exited the car. They had not washed any tires or administered any sanitation protocol. After their perfunctory introduction, they courteously apologized, "Oh, we didn't realize you had this going on." For all they knew it was a Sunday School class potluck.

To which I replied curtly, "That's the problem with you people. You never know what's going on." By this time, our seminar participants had all gathered around to watch the showdown on the Polyface front yard. At least on this day on this farm at this shakedown, it was about 40 to 2. Good guys on top. Bad guys seriously outnumbered. And probably outgunned. For sure outstmarted.

Kind of like the first time Allan Nation asked me to speak at his Stockman Grass Farmer conference in Jackson, Mississippi. He asked the 400-plus attendees for a show of hands: "How many of you are from the north?" About 100 hands went up. "How many from the south?" Everybody else's hands went up. "I think we could whip you boys today," he laughed good-naturedly.

The veterinarians said they wanted to take blood samples to check for avian flu. I responded emphatically, "You are not welcome here. You may not exit the car. You are trespassing and I demand that you leave immediately."

They backed out of the yard, turned around, and left. Again, they never stopped to sanitize their tires or do any sanitation protocol as they headed down the road to the next farm. I deduced right then that if anything was spreading the virus, these task force members driving around from farm to farm were probably the number one culprit.

I fully expected them to return Monday morning with a warrant, but they never did come back. Which is one reason why I encourage people being harassed by these bureaucrats to not be cowed into compliance. We've been told by these bureaucrats that if everyone treated them like we did, they couldn't begin to do their jobs. Well then, let's all treat them this way and maybe they'll get so stressed out they will die of heart attacks. The government is out of money and can't replace them because the U.S. is too busy building empires around the world, so let them just die off by attrition and pretty soon we'll have freedom by default. Kind of like going 80 miles per hour on the beltway marked 55. Who do you stop for speeding when all 5 lanes are running 25 miles an hour over the limit? Besides, the longer I can tie them down picking on me, the less time they have to pick on someone else.

The year was 2002, and it was another regional avian influenza outbreak in the poultry industry. This wild strain, known as low pathogenic, is about as harmful to poultry as a mild cold is to humans. It does not affect the meat or people who eat the meat, according to the USDA. I will say some things now that are conjecture because deep internal industry secrets are hard to uncover but I will lay down the information I have as clearly as possible to show you why many, many people here—including many industry growers—believe this was a contrived outbreak.

Here are the facts as I've been able to find them. The first flock in which the flu was found was not destroyed for several weeks. The flocks were not destroyed in the order in which the virus was found. The earliest flocks in which it was found were left alive for some time while others were destroyed immediately after discovery.

A couple of growers came to see me and reported that none of their birds exhibited any signs of sickness. Indeed, they reported that the birds were the healthiest they had ever raised. Because this strain is similar to a mild cold in humans, in a couple of days, any visible signs are gone and the birds go about their business.

By and large, the outbreak affected only the birds of the company that held the largest contract with Russian markets. Even though competitors' houses were much closer, the virus hopscotched to infect only the birds owned by the company that exported to Russian markets. Why didn't birds owned by the other growers become infected?

The outbreak occurred the day after the Russian markets collapsed due to allegations of dirty American chicken, nearly all of which came from one or two companies headquartered in Virginia's Shenandoah Valley. Within hours of the Russian import ban being lifted, the outbreak ceased.

The USDA sent scores of federal veterinarians to the area in roughly one-month rotations during the six-month outbreak. Official reports consistently blamed outdoor poultry and wild birds. Officials trapped wild birds and examined scores of backyard flocks and could never find one single carrier.

All in all, 1,000 tractor trailer loads of poultry were incinerated, landfilled, or buried on farms during that outbreak. The state and federal government indemnified the owners of the birds for their horrible losses. One catch—the industry owns the birds. The farmers just provide a house. Yes, the industry received the millions of indemnification dollars. Not the farmers.

Months after the outbreak, the only epidemiologically significant commonality as a vector to spread the virus was a rendering outfit north of Harrisonburg. All of the official press releases blaming backyard flocks and wild birds were contrived. And not true. The only common denominator actually quantified was the dump site at the disposal facility where farmers took their dead carcasses. Farmers who disposed of their mortalities on their own farms, either through incineration, burial, or composting, tended to escape the outbreak.

Here is what I believe happened, and many, many people on the inside of the industry agree with this—they are the ones who put two and two together to come up with this scenario. Like I said before, this may not be true, but it sure fits. And plenty of people much closer to the inside than I have said this to me.

Imagine yourself as a middle manager in the industry. A major foreign buyer announces a ban on your poultry. Your freezers are full. I'm talking about freezers that tractor trailers drive into. These are humongous drive-in freezer warehouses. You have a pipeline of birds heading to processing. You have breeder flocks out there producing eggs for the hatchery. You can't stick a cork in them with the command, "Shut down your eggs for a few weeks, hens. We'll tell you when we want some more."

The eggs keep coming. The chicks and poults keep coming. You have to make room in the warehouse for more birds, but there isn't any room. It's too costly from a public relations standpoint to throw them away. Where would you throw them? If somebody sees you dumping tractor trailer loads of frozen poultry, what will you say? What will happen to the value of your stocks?

Finally, you hit upon a plan, a wonderfully Grinchy kind of plan. A tiny vial of a fairly innocuous virus, a small outbreak, enough dead birds to get some breathing room, and the best part—the taxpayers indemnify the loss. Now don't get me wrong—I'm not saying this is what happened. I'm just saying that a lot of people inside the industry believe this is exactly what happened. And this is why it coincided so closely with the Russian market collapse.

The problem was that the outbreak became a little bigger than planned. Now it certainly may be that a conscious contamination did not occur. It could even be that some birds had it all along but it was kept hush-hush until orders came down from above that rather than sweeping the next positive-testing bird under the rug, it would be publicized. I don't know how it all played out internally. But virtually everyone in the industry believes a pile of hanky panky was going on somewhere.

During the outbreak, two of the federal vets came to visit us during their time off. They were not in any official capacity; they just wanted to see the farm that they'd heard about or read about. I was glad to have them. One came one week and the other came the following week. They did not know each other because they were from different parts of the country. But each of them said the same thing:

"Every one of us knows that the reason for the outbreak is too many birds in too tight living quarters in too many houses in too close a geographic proximity. But if any of us breathes a word of that publicly, we will be fired within 24 hours."

Now how does that make you feel about government protection? About the USDA scientists being the repository of food safety? About the integrity of their bulletins and reports?

Let me just say this as clearly as I know how: It's time for all of us to understand that no matter what the USDA press release says, no matter how official it looks, no matter how many scientists signed onto it, no matter the length of alphabet soup behind their names: the official finding is a lie.

Whether it's alleged mad sheep with Linda Fallaice in Vermont, the recall of raw milk from Mark MacAfee's Organic

Pastures Dairy in California, or avian flu or mad cow or hoof and mouth, the official government pronouncements are not the truth. Period.

We have an incredibly duplicitous culture today, trained like good little girls and boys to serve the state. In our schools we learned that the only questions worth asking are the ones the teacher deems important. The only subjects worth studying are the ones the guidance counselors recommend. The only curriculum worth pursuing is the one the accreditation board deems appropriate. And the only words worth saying are the ones that agree with the credentialed experts. If you don't believe that, just try taking a Biblical six-day creationist view in biology class and see where it gets you. We have become like sheep, and we're being sacrificed en masse to the interests that do not speak the truth, do not know the truth, and do not want to find the truth.

The truth will always be in the minority. Always, always, always. Jesus' principle of the broad way that leads to destruction and the narrow way that leads to life eternal (truth) spans every facet of life. Want to find the truth? You will never find it in the hallways of conventional institutions. You will find it in pockets, in clusters, individually. And it will not be mainstream.

Teddy Roosevelt used to say that nothing in government happens by accident. There is always an agenda. And especially today, the agenda usually involves more power and money to large corporate and bureaucratic interests with a parallel disempowering and impoverishing of smaller public and private entities.

Much evidence suggests that Prohibition was actually financed by powerful early petroleum interests to shut down the healthy, decentralized on-farm alcohol industry. I alluded to this in the previous chapter. The Model-T Ford had a carburetor adjustment and a dashboard adjustment so that it could run on either gasoline or alcohol. At that time, many, many farms made alcohol from corn, apples, sugarbeets and other starchy plants. It was a preservative, a disinfectant, a sedative, and a way to transport nutrient density in the days when transportation was slower and costlier. And before refrigeration.

Some alternative energy experts actually believe we would never have developed a dependency on foreign oil had we maintained the freedom for farmers to continue brewing alcohols to fuel the burgeoning automobile industry. But in the decade during Prohibition and the subsequent creation of the Bureau of Alcohol, Tobacco and Firearms, the domestic on-farm fuel industry was outlawed and destroyed, paving the way for the petroleum barons to dominate the energy scene. The right-wing prohibitionists were simply stool pigeons—albeit sincere of heart—for a sinister agenda to criminalize a decentralized energy system. And you thought it just happened.

The longer I live, the more amazed I am to discover that few things in history were in fact the way we learned them in school. And now along comes the book <u>1491</u> to explode all the thinking about manifest destiny, the American wilderness, and natural environments. Wow! What an exciting time to be alive.

I asked these two veterinarians, who were sympathetic to our pastured poultry operation, about a response protocol. See, I don't trust the laboratory. One independent poultry farmer in the county lost his entire operation based on one false positive. He tried desperately to protest. He quoted one official as telling him, "Look, I was sent to the Shenandoah Valley to kill chickens, and that's exactly what I'm going to do." Forget good science. Forget reason.

The farmer told me his flock was laying better than any he'd ever had in 40 years. They appeared healthy in every way. The vets just kept coming back taking samples and taking samples until the lab was able to concoct a reading they liked. He certainly didn't trust the lab, or the samples.

One lady here in Virginia was accused of having tainted cheese. The subsequent finding showed that the lab technician had dropped the sample on the floor, but rather than discarding it, just continued with the test. Those of you who think we need regulatory oversight for so many things, I have a question: why in the world would you think USDA bureaucrats are any more trustworthy or honorable than Pentagon bureaucrats?

Why the liberals who crucified Donald Rumsfeld, George W. Bush, and the concocted Weapons of Mass Destruction story against Iraq suddenly believe the USDA bureaucrats when they officially pontificate on something is beyond me. And why the conservatives who don't trust anything the education or social departments say would suddenly believe the USDA speaks truth is beyond me. I try to take a consistent approach: don't believe any of them. They all have unspoken agendas; they all want to grow their bureaucracy; their future depends on perpetuating problems,

We must be eclectic toward finding the truth. Read both Mother Jones and the Wall Street Journal. Listen to Jesse Jackson and Rush Limbaugh.

The two veterinarians who sat in the foyer of our house had some chilling words for me. Remember, these guys were sympathetic to our farm. They were friendly. They were not here to intimidate or threaten. "You have to cooperate anyway," they said.

"Why?"

"Because if you don't, you will be vilified as a Typhoid Mary. Think about it. All the neighbors submit to a blood sample, and you won't. What are they going to think? They will not think the lab is untrustworthy, like you do. Rather, they will think you have something to hide."

In that moment, I realized how desperate our situation has become. If I'm a thorn in their side, they will get me. It's like walking around with a big star of David emblazoned on my shirt in a Jewish ghetto in Warsaw in 1942. If I don't submit to whatever the officials want, the community views me as a rebellious troublemaker, a societal misfit. If I do submit, they make the tests show whatever they want so they can destroy me. When I watched the gut-wrenching movie The Pianist, I could identify with the desperation of the Jews. My heart ached for them, just like it aches today for all the local food farmers terrorized by the USDA.

How do you punch through that? I don't know. I just don't know. But I don't lie awake at night worrying about it. It's just the way it is. My response is to write this book, hoping to awaken a new generation to the machinations of government. And if these are just the stories of one little farmer in one little corner of

Virginia, imagine how many other stories could be told. Many of them are far more dramatic and draconian than mine. Most farmers who have tried what we have and run into these regulatory obstacles quietly go out of business. Very few farmers have the will or savvy to go up against these government agents. I have watched many, many farmers simply quit.

After having heard my personal story about bird flu, do you trust the official pronouncements about the more virulent Asian strains and the need to stockpile mountains of flu vaccine as a precautionary measure? Let's think this through just a minute. We know avian flu has been around since the late 1800s. It comes and goes.

But these allegedly new virulent strains arising out of the orient, what about them? Realize that intensive confinement poultry production must go hand-in-hand with an antibiotic regimen. In Asia, because of rampant thievery, poultry owners keep their birds inside or adjacent to their house. Otherwise, they would all be stolen. You folks who don't like our Judeao-Christian heritage should visit some of these other cultures just to see the rampant stealing and bribing that goes on.

On Asian farms, carbon is used for cooking and heating. Seldom, if ever, does anyone have sawdust or wood chips, or straw, to use as bedding. It just doesn't exist. Consequently, the typical Oriental backyard flock, which accounts for 75 percent of the production, consists of 100 birds in a tiny cage without any bedding. They stay there all the time under lock and key. Often the birds stay in the house at night, even roosting on the kitchen sink. For the record, chickens don't potty train very well.

Just because a production model is small does not make it hygienic. Some of the most filthy, unsanitary livestock production models in the world are the smallest. The two pigs kept in squalor out behind the house. The horse on the denuded paddock behind the farmette McMansion. The cow standing in the mud by the back gate. Nothing about smallness makes it inherently cleaner.

In the late 1950s when our family began farming in Venezuela, Dad quickly cornered the local poultry market because our chickens did not have subclinical pneumonia, indicated by mucous dripping down their beaks. In those Latin American open air markets, the farmers would bring their wares and vendors who had routes through the city would come and buy. The vendors were always looking for better quality merchandise because that is how they built their reputation with the homeowners along their delivery route.

With poultry, they would run their finger along the beak to feel the degree of mucous discharge. The drier the beak, the healthier the birds. The customers along the route of course would perform the same test. When they found one they liked, they would buy it off the vendor's shoulder pole, slaughter it and cook it for supper. When Dad began taking his chickens to the market, their beaks were dry. Vendors quickly learned and would line up at his stand, waiting for him to arrive.

Do you think the other farmers came to Dad asking him how he did it? No, they resented him for taking away their sales. Of course, he was keeping the birds on clean ground and providing clean water and a clean feeder. Quite simple, but substantially different than the indigenous approach.

What has happened in these foreign countries is that the rising economy and access to cheap subsidized grain has enabled these backyard poultry flocks to expand in recent years. The USDA has encouraged industrial farming in these countries to stimulate grain exports. In his New York Times bestseller Omnivore's Dilemma, Michael Pollan does a masterful job of describing the U.S. corn culture. The height, depth, and breadth of corn production is beyond imagination. It permeates every decision at the USDA and even foreign policy decisions.

One of my favorite stories in this regard comes from Allan Nation, editor of Stockman Grass Farmer magazine, who took a group of American grass farmers to Argentina to tour the world's best grass-based farms and ranches. When the American ambassador found out a group of American farmers and ranchers was in town, he invited them over to the Embassy for refreshments.

"What are you all doing in Argentina?" he asked.

"We're here to study grass-based agriculture and see firsthand how to finish beef on grass instead of in feedlots," replied Nation.

At that point, Nation said the ambassador nearly kicked them out of the embassy, proclaiming, "Don't you understand why I am here? I am here to make Argentina quit grass-finishing beef and begin using corn in a feedlot." There's the good ol' USA, friend of environmental agriculture, respecter of other cultures. Yessireee.

Too many people really think the world revolves around Broadway, Hollywood, and Nashville. It doesn't—even as big and influential as they are. Corn dominates the food industry, the machinery industry, the petroleum industry (fertilizer), the chemical industry, the genetic and seed industry, and the livestock industry. As corn goes, perhaps we could say, so goes the nation. No matter what happens, corn production must be encouraged both domestically and internationally in order to keep Wall Street afloat. Our entire culture floats on corn, believe it or not.

Would that it floated on grass, but that would require a total realignment of power and money. Adjustments on this kind of scale are culturally cataclysmic. I say bring it on. It's long overdue. With the ready access to additional corn, these backyard flocks expand beyond their carrying capacity. A pen that used to hold 20 birds now holds 100.

Staying on a foot of pure droppings, without bedding, and tracking their feces through their water pans creates terribly unsanitary conditions for the poultry. The owners can't afford the antibiotics that industrialized nations would use in similar circumstances. Add to that the international trafficking in chicken parts from industrialized outfits, along with overuse of antibiotics that create super-bugs, and the whole system becomes a conflagration of stressed immune systems.

The spread of the newly virulent avian flu follows the transportation paths of chicken parts and the heavy industrialization of poultry production. Michael Greger has written a fantastic book tracing the roots of bird flu: Bird Flu: A Virus of Our Own Hatching. He documents clearly that the only way the virulent, high pathogen strains could mutate and jump to domestic fowl was through industrialized poultry farming that crowded chickens close enough to feed the virus. Throughout the millennia, these strains have never existed because domestic poultry was not raised in such immuno-compromised conditions. This book is impeccably researched yet easy to read.

Many people don't realize that feathers, blood, and guts are transported internationally as cheap protein supplements for animal feeds. An examination of the spread certainly indicates these traffic routes as a vector. Possibly some of it is spread by wild birds. But waterfowl are fairly immune to it. British research suggests that some of this immunity may be due to grass consumption—waterfowl eat way more grass than other poultry.

At any rate, the international community makes no differentiation between clean and unclean outdoor poultry. Make no mistake, the industry labels farmers like us bioterrorists because our outdoor poultry commiserates with Red-Winged Blackbirds and sparrows, allegedly transporting viruses to the scientifically based industrial poultry houses. That's the official line.

If that's the case, why, in 2002 when 1,000 tractor trailer loads of industrial poultry were destroyed in this area, could not one outdoor bird—domestic or wild—be found with the virus? Not one? And yet all during the outbreak, every non-industrial poultry producer was demonized as jeopardizing the industry. In fact, even farm ponds were identified as a liability. No science; just fear factoring.

When people are afraid, they don't think. They simply react. And they are willing to give up reason, give up freedom, give up everything to feel protected.

When the virulent flu showed up in northern India in early 2006, more than1 million birds were destroyed in a radius of 1.5 miles of the outbreak. Folks, that's not backyard poultry. That is intensive industrial-duplicated production on a magnitude that would make Tyson envious. Suffice it to say that, again, the USDA findings and the protocol for dealing with an outbreak are completely off base. They will not do anything to diminish the power, reputation, and sales of the industrial poultry sector. That we can all count on.

The very notion that chickens allowed access to the outdoors threaten human health is just inherently ridiculous. Civilizations have been raising outdoor poultry for how long? Does anyone stop to think for a moment what is there about poultry production in the last couple of decades that is different from historic models that could have created this new disease? Hello? Could it be crowding on a scale we've never seen before?

What is different is global parts trafficking, intensification of filthy squalor production models, antibiotic-resistant mutated strains, and industrial mega-houses. But does any official have the courage to touch any of these sacred elements? No, they all line up behind the official verbiage that farmers like me are trying to kill nice folks like you.

In 2002, officials said that one feather carried enough virus to infect 500,000 birds. Have you driven down the interstate lately behind one of those tractor trailers hauling birds to the processing plant? It's like a snowstorm. Feathers by the millions line the road side ditches. On the other hand, our farm exemplifies true biosecurity because the chickens leave only in neat little coolers. We aren't seeding the community with feathers and feces.

The USDA makes its view abundantly clear in its official bulletins: diseases are like little fairies hovering over the landscape. Completely whimsical, they flit about playing roulette with the diseases in their arsenal. For no reason whatsoever, they pull a bit of avian flu dust out of their bag and sprinkle it on Farmer John's place, laughing as they go.

A couple of days later, they reach into another bag and scatter a smattering of mad cow on Farmer Bob's place. Twittering excitedly, they flit over a couple of counties, close their eyes, and then grab some hoof and mouth dust. The apparent conclusion of the experts is that we are all just sitting ducks, and these fairies zap diseases on our animals for no apparent reason and for no apparent cause. Farmers are simply at their mercy, and we can do nothing to increase immunity in our animals or protect our premises from disease.

How ridiculous. Of course we can strengthen our immune systems. Of course we can build up the health of our animals and plants. But the best immunological work the industry and government agents can imagine comes in little bottles from pharmaceutical companies. It never occurs to them to fundamentally change the production models. To quit feeding herbivores dead cows, dead chickens, chicken manure and grain, and to let them graze like herbivores again.

It never occurs to them to vacate the CAFO's and raise pastured poultry and hogs. It never occurs to them to shut down confinement dairies and let all those cows graze. Goodness, they never even think that deserts might be inappropriate places to produce the nation's milk. And they certainly never think about creating local food networks. That would never do.

Bottom line: normal avian flu does not hurt anybody, including the poultry. No massive killing is necessary. New virulent strains are a product of global trade, production intensity, lack of hygiene, and concoction-induced mutations. That's much closer to the truth than anything coming out of the USDA.

Chapter 20

Bioterrorism

Every time I hear a government official use the word bioterrorism, I know two things will follow:

1. No interest in addressing the most unsafe parts of the food system.

2. Whatever solution is offered will stifle the safest part of the food system.

How can I be so sure? Look at the track record. A perusal of any report on bioterrorism, whether generated by the private sector or the public sector, identifies three vulnerabilities of the American food system:

1. Centralized production.

2. Centralized processing.

3. Centralized warehousing, including long distance transportation.

But also in every single report I've seen, these vulnerabilities are just presented as inherent facts of life. In other words, no fundamental alternative exists to these weaknesses; they just are and nothing can be done to change that.

The result is that every single solution assumes some centralized bureaucratic regulatory protocol. The proposals for policing the current food system would make any freedom loving or mildly sensible person shudder.

Let's look at these issues one at a time. Centralized production means the farms are huge and the acreage is relatively devoid of people. Expanses of landscape just stretch to the horizon. If you drive up to the average Concentrated Animal Feeding Operation (CAFO) you will not find anyone around.

The confined animals are on automatic waterers and automatic feeders. Thermostats turn fans and heaters on and off as necessary. Drive up to a Midwest corn field and see how many farmers you see. These systems produce mountains of food but the owners are seldom around to see what's going on.

In highly-integrated smaller farms and production facilities, however, human activity is present. Neighbors see cars going down the road. Farmers can see their neighbors, and are aware of anything out of the ordinary. Smaller farms tend to be a little less automated, which requires more warm bodies to do the work. The loving warm bodies provide a security against strangers lurking around.

To help picture what I'm describing, imagine trying to sneak into an Amish farm in Lancaster County, Pennsylvania. The community would be abuzz with information about a strange person lurking about. This is the whole principle of the Neighborhood Watch program. Smaller community-based farms are populated with more eyes and ears—partly because these farms are more aesthetically and aromatically pleasant. CAFOs exude such a stench that no person wants to be around them anyway—including the farmers.

Centralized processing is the second vulnerability of the food system. Enough has been written on this issue that I don't need to rehash it here. Suffice it to say that if you walk into any large processing facility in the U.S., you will think you've entered a foreign country. And I'm not against foreigners. But in discussions of bioterrorism, the threat is always described as coming from a foreigner.

These facilities are swarming with people. In this case, we have the opposite scenario from the centralized farm. Here, the facilities have such a mob of people, many speaking different languages, that security is just as difficult but for the opposite reason. In one case, no one is around; in the other, too many people are around. Either extreme poses security risks.

Again, to put this in perspective, think of all the movies you've seen where the criminal melts into a crowd. The safest place to hide is in a mob. Where's Waldo? Why can't you find Waldo? Because he's in a crowd. People coming and going, mingling, jostling, taking bathroom breaks. It's a veritable madhouse. I've heard from people who work in these plants what happens when the INS (Immigration and Naturalization Service) agents show up for a raid.

I'd love to have a video. Workers hide in dumpsters. They jump into trash cans. They climb in behind the massive refrigeration units. They crawl under meat buggies and climb up into the ceiling above the meat rails.

Compare that with a facility of fewer than 50 people, where everybody knows everybody else's name. A new face only shows up once a week. The owner and manager lurk near enough to see everyone who enters and exits the plant. Only one or two people have a key. The facility is imbedded in the community. A plant with 50 workers is still plenty big enough to be efficient, but is small enough to be secure.

Our culture's love affair with size comes back to haunt us in all kinds of different ways. Inability to police the integrity of workers in the food processing industry is just one of those ways. Most of the smaller plants I'm familiar with pay higher salaries than the biggest plants. The worker salaries drop as the plant size increases. With lower salaries, fewer people come to work there from the neighborhood. Larger plants draw their workers from farther afield, reducing the ability to check out backgrounds and criminal records.

Besides, if a bioterrorist wants to act, more people can be harmed by tainting 10 tons of food rather than only 1 ton. A plant slaughtering 5,000 beeves per day must move 45 tractor trailer loads of finished product out of the plant per day. Every machine, every grinder, every conveyor is supersized in such a facility. That's where the potential terrorist will more easily get a job, blend in, and slip the vial of poison into the food.

Finally, centralized warehousing and long distance transportation. The food distribution system in this country is practically beyond imagination. Just hauling French fries to McDonald's restaurants is a massive, massive materials handling exercise. I've been in a couple of food warehouses. They are indescribably large. Pallets on racks clear up to 40 ft. ceilings.

Boxes of food. Tons of food. Millions of prepared dinners. And they sit, and sit, and sit. The pipeline to keep such a massive food chain operating requires gigantic inventories to protect against Teamster strikes, drought, or blizzards. These billions of boxes move by truck and train over thousands of miles of roads, handled by all sorts of people.

Just for perspective, think about the vulnerability of a pastured broiler from our farm. We purchase the chicks from a hatchery and unload them into our brooder house. It sits 50 yards from the main farmhouse where Teresa and I live, 50 yards from Mom's house, and 50 yards from the apprentice cottage. It also sits about 75 yards from the sales building, to encourage customers to go out and take a look at the little balls of fluff—especially customers with small children.

After a couple of weeks we take them out to our fields, where we move them every morning and check on them every evening. At night, we are all sleeping within a few hundred yards of them. Our neighbors are all aware if strange cars come down our road.

When harvest day comes, we load them up in crates and bring them into the processing shed, located 20 yards behind our house and abutting the sales building. This all literally happens in the back yard. A crew of 8-12 works together for a couple of hours, then we package them and put them in the freezer.

The walk-in freezer is located 30 yards from our back door and adjacent to the sales building and customer parking lot. In order to get there, a person has to come in our lane, in front of our house, in front of mom's house, and in view of the apprentice cottage and Daniel and Sheri's house. When a customer in Reston orders a chicken from us, to be delivered at our next scheduled drop, we take the frozen chicken out of the walk-in and put it in a cooler. We put the cooler back in the walk-in the night before the delivery and take it up to Reston in our delivery bus.

The customer meets us at the rendezvous site, picks up the chicken, and takes it home to her freezer. That chicken was literally never out of our sight from arrival as a hatchling to final customer. What are the chances for someone to terrorize that food? Virtually nil.

And that's the point of this whole discussion. Unless and until government policy encourages a local food chain, America's food system will be increasingly vulnerable to bioterrorism.

But with every new announcement comes increased demands, especially from the consumer advocacy sector, for more regulations. These additional regulations simply destroy the decentralized portions of the food system because only the centralized elements can handle the overhead to fill out more paperwork. Software to process the forms costs the same whether it's handling 2,000 pages or 1. The ink is not expensive. Neither is the electronic impulse to send the message. It's entering and processing the required data to prove that my business has complied with all the firewalls that makes small outfit compliance economically and logistically impossible.

If you ask me, the biggest bioterrorist threat to the food system comes from pin-striped suits sitting at big oval tables on over-stuffed chairs inside the Beltway. I fear the politicians and bureaucrats far more than some foreign terrorist. The foreign terrorists don't know about my little farm and my little flock of chickens and my little group of customers. We're just a fly on the wall, unobservable to groups that would seek to harm Americans.

Besides, the foreigners don't set domestic policy. They have to penetrate the system from outside. But these domestic terrorists in pin-striped suits work from the inside. They make a phone call to my local sheriff, the sheriff sends the deputies, and my goose is cooked. My outside-the-system food chain scares them to death, because it's hard to control.

When fear takes hold, everybody, including the politicians, starts trying to get control. Free spirits like me represent a lack of patriotic fervor. In the name of patriotism, we must all pull together, be united. Quash those who would dare say and do differently. What's amazing is that the least vulnerable parts of the food system—like our farm and our processing and our customers—are the most vulnerable to government policy. The very protective policy that is supposed to reduce vulnerability.

It's insane. Just so we're all clear about what would actually create a decentralized production, processing, and warehousing food system, let me throw some ideas on the table. These ideas are fluid; they aren't carved in stone. This is just the kind of discussion I'd like to see somewhere—anywhere. I've tried to testify on the Farm Bills for nearly 20 years and talked to senators and congressman, both sides of the aisle, who said they'd get me in. But it never happens. These issues do not even merit the light of day in Washington. Lacking any other forum for bringing them to the public, therefore, I will now use this book to introduce them to the public.

1. Eliminate ALL agricultural production subsidies. All of the subsidies target certain commodities, and by doing so create overabundance of those items while creating a climate of competitive disadvantage to alternative items. Farming should stand on its own; as soon as it must, only the type that can, will.

2. Eliminate all grants and tax concessions to any private business for anything. This includes airlines, automakers, ethanol plants, and processing facilities.

3. Eliminate funding to land grant colleges and universities. These public institutions provide the academic leverage to corporations to not only pay for the research private businesses should pay for as part of their R&D program, but lends credentialed prejudicial support to discoveries and models that

would never become widely used otherwise. In other words, if we farmer greenies could argue toe-to-toe with the industrial food system, we could win. But we must also argue with the credentialed academic land grant college community. It's two against one; a stacked deck.

4. Allow any citizen to waive their right to government-sanctioned food. A universal opt-out mechanism to self-protect and self-inspect. The waiver would include giving up the right to sue the food seller for anything. Period. That protects the farmer and shares the responsibility. This is the number one reason given by bureaucrats to justify allowing only government-inspected food on a dinner plate—to protect the food industry from being sued by consumers. This really trumps the idea of safe food.

5. Anyone may grow, process, and deliver any food directly to the end user. End users include restaurants, institutions, and individual customers for home preparation. This includes dairy products, milk, meat, and poultry.

6. Institute complaint-driven inspection with unannounced sampling from the retail package. As long as the food is clean, it's clean. Doesn't matter how it got that way. Get the inspectors out of the plants and let that be privatized to consumer groups like AAA or Underwriter's Laboratories. The whole inspection bureaucracy would then be dismantled and efficacy-driven.

That's enough for starters. Essentially, this little set of conversation-starters gets rid of the economic engine that drives continued centralization. It frees up farmers and home kitchens to freely access their neighborhood with local food. The combination of creating freedom from the bottom up and eliminating improper incentives from the top down would fundamentally alter the food system. Thousands of entrepreneurial farmers and cottage industries would spring up to service their neighborhoods.

Any fear that big companies would become bigger and/or worse is countered by the fact that thousands of little competitors will spring up. The big guys will have to compete on the basis of truth, not slick PR and political clout. This democratizes the food system, which creates its own checks and balances.

I realize this agenda seems incredibly radical. All real solutions are revolutionary. And if people are afraid of doing this on a nationwide scale, how about a couple of localities offering it as a prototype? For you people who think dirty farmers would kill their neighbors with dirty milk, how about trying it somewhere? Right now, even if a community tried this the feds would be on them like fleas on a dog. "You can't do that. You can't do that."

Every time I've testified for any concessions in Virginia, that's the immediate cry. "We'll withdraw federal funding from your schools and roads." It's ridiculous. What's wrong with prototyping a revolutionary idea somewhere? If the citizens in that community are willing, why not let them self-discover? It can't cause any more harm than the current system.

Any abuses are limited to that little area. I think with all the hoopla over Enterprise Zones we need some Food System Innovation Zones. I don't care if it takes awhile to prove the soundness of these ideas. I don't even care if this is too radical. But we should at least be willing to let someone somewhere try. If it fails it fails. But I have a sneaking suspicion it would be a resounding success. And that is what the powers that be fear---the success of the truth.

Chapter 21

National Animal Identification System

The National Animal Identification System (NAIS) is the latest greatest effort by industry to use the government to placate the peasants. As the grassroots backlash builds, this program has been referred to as the "mark of the beast" from quarters that don't normally invoke Biblical references.

This is the hardest chapter of this book to write because NAIS is in such a fluid state as I write this in early January, 2007. The NO-NAIS camp is certainly gaining strength. But then again the other side is extremely entrenched, and I don't have much faith that enough American people will understand the issue—or even care about it. Unless this book gets on Oprah.

For the uninformed, NAIS is a proposal, already mandatory in Wisconsin and Indiana, to put Radio Frequency Identification (RFID) chips in every animal—cows, pigs, chickens, horses, sheep, goats. In the earliest plans, it included every fish. Millions and millions of dollars have already been spent on this plan that was initially to be mandatory by July 1, 2007. That has now been changed to sometime in 2009. Now many government officials are saying it will never be mandatory. But even a voluntary system will have what the USDA calls "chokeholds" that virtually make it mandatory.

Essentially, the system contains three parts:

1. Premises Identification. Every single piece of property that domiciles an animal is supposed to register with the USDA. According to government officials, this enables "us to find you" in case a disease outbreak occurs. The ostensible goal is to protect farmers from an outbreak. Kind of like knowing which houses are occupied so that in an emergency everyone can be evacuated and no one will be missed.

2. Individual Animal Identification. Installing RFID chips and entering that information in a database would allow each animal's existence to be known. In other words, if an animal exists, it is in the database.

3. Trace back and Tracking. The final step is to enter every animal's whereabouts, coupled with the premises identification, into the database. That way if any animal is sick with a communicable disease, within 24 hours its entire history can be ascertained and every animal it ever came in contact with could be isolated. In the case of beef cattle, the average steer lives on several different places. It's born on a cow-calf farm, then weaned and sold to a stocker operation, then sold to a finishing operation—feedlot.

To work, the system requires every movement of every animal to be entered into a database. Under U.S. code each noncompliant event carries a fine of up to $1,000 per infraction, per day:

- Selling an unregistered animal.
- Buying an unregistered animal.

- Moving an animal without database notification—including taking your horse on a trail ride beyond the confines of its registered premise of domicile.
- Owning animals that have come from somewhere else that were or are unregistered.
- Failure to notify the database within 24 hours of any change of location, including if you move a cow from one rental farm to another rental farm; including if the cows accidentally wandered into the neighbor's field.
- Failure to register an animal's death; failure to delete a nonliving animal from the database.

If at this point you are thinking totalitarianism, you are right. Plenty of conspiracy theorists see this system as the prototype for tracking people. This is the great equal and opposite reaction to technology. The innovation that allows security tracking for good purposes like finding lost children enables evil intentions to move forward with sinister agendas.

At the risk of sounding completely naïve, I will say flat out that I do not sense a conspiracy behind this effort. Like so many other government plans, this is another well-intentioned idea with incredibly terrible consequences. I happen to believe that when the food police say raw milk will kill you, they actually believe it. When they say my outdoor chickens are dangerous to the Tyson houses, they sincerely believe it. When they say consumers are too stupid to make food choices, they actually believe it. I think people sincerely believe a lot of stupid things. This is just stupidity, not conspiracy. It's a fraternity of stupidity, to be sure, but we need not label it a conspiracy.

Just to put NAIS in perspective, realize that for the first time in human history, this will require a license (registration) from the government in order for one person to have one chicken for a breakfast egg for their own table, assuming that chicken was purchased from someone. That is a policy with far-reaching consequences. And that such a policy has moved down the political process this far without even 10 percent of Americans being aware of it should give us all pause. The USDA has done everything possible to keep this plan under wraps and spring it on an unsuspecting populace.

This is such a huge topic I hardly know where to start. For those of you reading about this for the first time, you're probably in shock and could use some humor. For those of you who have been faithfully fighting this plan, you're probably raging by now and could use some humor as well. For the sake of our collective sanity, then, let's assume that all of this goes apace and the year is 2011.

A car with blue government plates rolls up to our front gate and a government NAIS auditor steps out, flashing his big bronze badge.

"I've come to audit your poultry for NAIS compliance."

"Okay, dear government friend, let's go out to the Eggmobile and take a look." I run into the house and retrieve my computer printout of the 1,000 RFID chip numbers on that flock of laying hens.

We walk out to the field and the auditor fires up his handy-dandy monitoring baton and it starts beeping. He plugs it into a printer dangling from his belt and a tape of numbers begins spewing from the printer. All of this, of course, assumes that none of the implanted chips has failed and that the monitor has no metallic or other-wave interference. Everything works perfectly, like most human inventions do.

I begin matching the numbers to my in-house list. Let's see, 540279632, 540279638, 540279691, 540279820 . . .

An hour later, the monitor quits beeping and he looks at the total. "999. Hmmm, Mr. Salatin, you are not compliant One chicken is missing. Please find the other chicken. I will issue you a summons to appear in court four weeks from today. The fine is only $500 for this infraction. I will return in two weeks to verify compliance. Good day, sir."

Now I'm going nuts. What happened to that other chicken? I call everyone together and we begin canvassing the areas where the Eggmobile has been. Let's see, yesterday it was up in the far flat field. Two days before that it was in the beehive field. By afternoon, we've found the pile of feathers. Clearly a hawk attack. Hawks only eat the heads—where the RFID chip is. Buzzards got the rest.

We spend the next two days trying to trap the hawk in a live trap. This activity, of course, is highly illegal, but we must find that chip. Finally we trap the hawk, put him in a cage, and place him right inside the Eggmobile to await the auditor's return.

As scheduled, he shows up two weeks later and we trudge to the field. We begin the whole process again. The hawk is positioned discreetly behind the Eggmobile door. We simply stand in the field to do our inventory. After two hours verifying all the numbers, everything matches perfectly. The auditor turns to me: "Very good, sir. I'm glad you found the chicken. You have passed the compliance test. Good day."

Now we're compliant, but we have this endangered species in a cage. "You know, this is going to be a pain having to catch field mice for this hawk to keep it alive. What do we do now?"

One of the apprentices who graduated from a university with an environmental sciences degree offers the solution: "Why don't we kill the hawk, remove the chip, shove it into another chicken so it has two? These officials don't count anything; all they care about is the chips. They won't know a bird is missing as long as all the chips beep in."

Great idea. We proceed to do it. A few days later we go to court, pay the fine, raise the price of our eggs 20 cents a dozen, and live happily ever after. The end.

Can you imagine farming like this? The ability to abuse the power, to move at whim against this outfit or that outfit, are simply too obvious to deny. Now let's dissect this issue one step at a time.

Has it been tried anywhere? Do we have any prototype anywhere? Kind of. Right now, two countries have implemented the program, but only for cattle. Australia and Canada. Australia's four-year record is dismal. First of all, the equipment that reads the chips averages a 3-7 percent errancy rate. In other words, if they run 100 steers through a chute at a cattle auction, 3-

7 of those animals cannot be identified. That may mean the RFID chip failed or for whatever reason, the monitor can't pick up the signal. They've had big problems with interference from metal. Since most of the chutes are made from metal pipe, this is a major problem.

The monitoring delays have actually killed cattle. Auctions that used to take a day have taken up to three days. The stress on the animals from being penned up waiting to go through the monitoring device is serious—sometimes resulting in death. Each auction facility had to buy a $300,000 monitoring device, not a small investment for a business like this. The bottom line is that on average, Australia is running a one million head annual discrepancy rate.

In four years, four million animals cannot be found. Generally, they are not deleted from the database when they are slaughtered. When farmers ask for a printout of their numbers, animals that they sold for slaughter have not be deleted. All of us know how hard it is to get information out of a computer system. Computers are designed to save information, to be incredibly efficient accumulators of data. All sorts of firewalls exist to prevent the computer from losing information. We have this saying around here: "Once you're in that computer, forget it." If someone enters your address with a house number wrong, it will come that way until doomsday, no matter how many times you call or write and instruct the company that the number is wrong.

It's no wonder that the discrepancy rate is that high. But even with that dismal record, USDA officials keep telling the American public that this system is absolutely necessary to stop disease and will work. For crying out loud, how can a government that can't even keep up with a few million illegal aliens possibly keep up with billions of cows, chickens, turkeys, pigs, horses, sheep, and goats? And fish.

I'll tell you what. How about coming over to my house next Sunday? I have a souped-up go-cart with an acetylene torch bottle mounted on one side. I'll strap an oxygen bottle to us, you and me, and we'll light that thing off, fly to the moon and back, and have a great old time. Folks, just because I say something will happen doesn't mean it will. And just because the USDA says the NAIS will provide 24-hour trace-back and will prevent disease doesn't mean it will happen.

The Australia prototype, if it illustrates anything, illustrates the ineffectiveness of this idea, not the effectiveness. Any reasonable person, when looking at the Australian mess, would say, "Well, at least we know what won't work. Forget that idea. Let's try something totally different." But no, the best and brightest minds in American agriculture dismiss this dismal track record out of hand.

A large-scale cattle farmer and industry leader told me, "Whatever problems we have with the technology will be worked out as we go along." Such blind faith in the technology. The same type of people said the same thing about residual affects of DDT. They say the same things about the carcinogenic compounds generated by irradiation. They say the same thing about the devastating side effects already occurring through genetic engineering. The folks who think we're always clever enough to technology our way out of technological catastrophes are doomed. Nature bats last. I'm not a Luddite by any stretch, but neither am I just a blind technology worshipper. Somewhere we need a balance.

In Canada, the track record, again only on cattle, is equally abysmal. I spoke to several farmers up there this winter when I was doing some seminars, and they said the program was a joke. "These tags fall out and we just stick another one in, but nobody checks and nobody knows anything," they said. The truth is that it is a clever public relations campaign to create the semblance of protection for an ignorant, duplicitous public.

And now Canadian cattle have been found in American sale barns. These are cattle with all sorts of prohibitions on import because of mad cow found in Canadian beef, yet they are not being screened at the border and clearly their records are either being falsified or not checked. Think about human identity theft. It's a huge issue. And it's an issue on people who have driver's licenses, social insecurity cards, passports. Now imagine what this would be like on way more animals, and on critters who have none of this other paperwork. Talk about a nightmare.

The Canadian experience is similar to what I hear from a farmer friend in China, regarding all the World Health Organization avian influenza programs. All those media reports about vaccinations and new protocols being faithfully carried out to stop the spread of AI are false. "It's all paperwork. The local officials aren't visiting anybody's chickens and nothing is happening. They fill out forms, send them to the WHO, which compiles them into credible-sounding programs that merit press releases, and it's all a charade." That's the official story from a farmer on the ground in the village. The older I get, the less surprised I am at these realities. And the less I trust anything from official sources. Nearly every statement from on high is just some politically-massaged spin to obfuscate the truth. And placate the peasants.

The Canadian farmers told me that due to the problem they were having with RFID ear tags, some officials were pushing for subcutaneous chips (imbedded under the skin). But even the industry is balking at that notion, because the slaughter houses don't want the burden of finding them and removing them. Remember, these things are the size of a large pepper flake. And when implanted, they sometimes dislodge and flow in the bloodstream to other parts of the body. Can you imagine having to find and verify removal of every chip on every carcass before it could move on down the line to the chill room?

Consumer advocacy groups, bless their heart, have made it clear that people don't want to eat stray RFID chips. We could become walking transmitters. Would these things move around in our bloodstream and block an artery? These are not silly questions. And that this program has consumed the millions of dollars it has, and moved as far down the road as it has, with no more definitive answers to some of these questions than we have, is truly remarkable.

I spoke with an attorney who represents one of the largest food businesses in the world. He came to the farm for a visit, a delightful gentleman, and we had an amicable wide-ranging discussion. When NAIS came up, he said he would put it to me straight, "People don't trust the large corporations. If you're a large corporation, you need that trust to survive. How do you get that trust? You create a system that makes it look like you care. People want to see you doing something that protects them. That is how the NAIS program came to be. But, and here's the other part of the equation, if you're the chief executive of a large business, you don't want to pay for it. Instead, you wine and dine politicians to convince them that they will curry favor with their constituents if they demand this program. Now you have people's faith without having to pay for it."

I was shocked by the matter-of-fact way this high level attorney spelled out what, for him, is just something else in a day's business. That high rollers inside the beltway just meet for coffee and connive such plots is more amazing than fiction. That also helps to explain why the plan, like all government plans, heavily discriminates against small producers.

For example, chickens and turkeys owned by vertically integrated industries like Tyson or Pilgrim's Pride would only be assessed one number per flock. Each factory house flock would only be assigned one number. That's one number per, say, 10,000 birds, whereas a small operator like us would need a number for every bird—10,000 numbers. It doesn't take a rocket scientist to see quickly that such a system creates an undo burden on small operators.

But here's an interesting permutation. Suppose one of those industrial chickens—and this occurs all the time—happens to fall off one of the tractor trailers hurtling up the interstate hauling chickens, coating the side ditches with virally-infected feathers? And suppose one of those chickens—and it occurs all the time—happens to wander into my backyard and join my chickens? And suppose I don't notice it right away, and the chicken actually gets healthy and starts looking like my chickens? And then the auditor comes. I've got an alleged bootleg, black market chicken. I face the sheriff, then the judge, and nobody believes that I don't know where the unregistered chicken came from.

To show the magnitude of all this, realize that Los Angeles county alone has a reported 60,000 backyard flocks of chickens, most of which are owned by Asians and Hispanics, many of whom themselves lack some credentialed citizenship paperwork. I'm sure they will all ante up to the registration booth and get their premises ID'ed. Then I'm sure they will install RFID tags in their illegal fighting cocks and the other poultry. And I'm sure they will report every movement to the proper authorities. So what is our country going to do? Send in the Marines and round up all the chickens? Come on, get real.

Maybe they'll offer a concession, like a temporary chicken resident amnesty program. As long as the chicken is dead within two years of hatching, no registry is necessary.

Serious problems exist with this program. Not the least of which is the logistical infrastructure necessary for a farmer to comply. Certainly the data entry will be electronic, via the internet. What about Amish who don't have computers and whose beliefs forbid owning one? What about all of us over 50 who don't know how to get on the internet? Assuming the government penetrates religious freedoms and everyone gets their computer, what happens if mine crashes the day I happen to be moving cows from one farm to another? Or the day the goats happen to get onto the neighbor's place and munch her tulips? Or my daughter takes her horse to a 4-H trail ride 10 miles away?

My experience with bureaucrats is that they have never heard the word mercy. As one Amish man told me, "Bureaucrats have no heart." They just fill out the forms and do what they are told. My catastrophes and wrecks hold short shrift with these government agents. All they care about is if their paperwork is in order. And if you get in the way of their paperwork, there's hell to pay.

The biggest NAIS unknown, and most unclear, is the cost and who will bear the burden. One week it looks like the farmers will have to pay for all of it. The next week, the government. Trying to pin officials down on this most important aspect has been elusive in the extreme. RFID tags cost anywhere from 40 cents to $30. That's a pretty big deal when we're talking about chickens.

To his credit, Congressman Bob Goodlatte, (R-Va.) former chairman of the House Agriculture Committee, sent the USDA some 50 questions for clarification in the spring of 2006. In more than a year, even he has not received answers to those questions. If he doesn't know what's going on, I don't know how the rest of us hope to know. But the unknown doesn't slow down the NAIS juggernaut. Government agriculture organizations, at taxpayer expense, of course, are buying huge wrap-around ads on rural magazine covers touting the virtues of this system. It's all for your protection, of course.

Recently, the USDA has begun retreating from its word mandatory and moved to using the word voluntary. But it can be made de facto mandatory quite easily. Most of us who have been watching the program's evolution believe that the enforcement will be at certain choke points in the system.

For example, all agricultural fairs will have to record the number of each animal at the fair. This is the way for the politicians and bureaucrats to say the program is voluntary when it's really not. In order to exhibit an animal, the farmer would have to comply with the whole process.

Another huge choke point would be slaughter houses. With home slaughter prohibited both by inspection and zoning, farmers must take their animals to an abattoir. Quite easily a mandatory NAIS compliance requirement could be implemented on all slaughter houses. If your animal isn't fully compliant with the voluntary NAIS, you can't unload it.

With all this said, let's just imagine that the program went forward, the database was indeed established, and RFID chips went into the animals. Would it work? I mean, would it protect anyone from anything?

Realize that it doesn't do any more monitoring for disease. So few animals are actually checked for anything that the chances of finding anything are minimal. Most of the diseases that people are afraid of happen after animals leave the farm. Dirty slaughtering procedures can infect animals. What's the point of blaming that on the farmer?

The way I see it, the system cannot and will not work. It will simply become another politicized, bureaucratized centralized program to encourage a centralized food system. Whatever gets punished in the process will be parts that are not centralized.

I think the bigger question, and the one I'd like to ask the Secretary of Agriculture, is this: "Can farmers do anything to reduce the risk of disease? If farmers could do anything to reduce disease, what would it be?"

That question puts the official in a dilemma. If he says no then he's a fool. We all know, intuitively, that farmers can do things to reduce disease. He can't afford to say no because he understands this. It's kind of like asking someone: "Is there anything in your life that you could do better?" Of course he has to say yes.

When he says yes, the next question, of course, is, "What would that be?"

I don't know what the Secretary would say, but I'll throw out some ideas that might make a difference:

- Feed herbivores like herbivores—no silage, no grain, no chicken manure; just grass and hay.
- Increase floor size in confinement chicken houses to at least 3 square feet per chicken and 12 square feet per turkey, OR shut down the confinement houses and move to pastured poultry (not free range, but pastured—daily or extremely frequent complete moves to fresh pasture).
- Increase mineral feeding by eliminating mineral blocks (a cow or horse needs a tongue the size of an aircraft carrier to lick off enough to do any good) and offering granular, including kelp, and all they want.
- Community-based, appropriate-sized slaughter houses and processing facilities.
- No hormone, subtherapeutic antibiotic, systemic parasiticides.

- Eliminate chemical fertilizers.
- Compost all livestock manures—no burning for electricity generation; no lagoons and slurry systems.
- No farm could generate more manure than its land base can environmentally metabolize.
- No corn or soybeans grown more than two consecutive years on the same ground. Open pollinated corn rather than hybrid corn.

Lest I be misunderstood, let me say this emphatically: This is not an agenda that should be mandated. It is an agenda that the culture could endorse, and if we still felt like we needed a Secretary of Agriculture (and that certainly is a debatable point) that position would use its power to promote these kinds of production systems. Can you imagine if the Secretary of Agriculture used that bully pulpit to promote such an agenda?

The reason NAIS can't help us is because the USDA attitude toward disease is simply to annihilate. Whether it's poultry, sheep, or cows, the response is simply to kill everything. Every time we've had the tiniest avian influenza outbreak, the only solution is depopulation, which is a euphemistic way to say: "Kill everything." If one bird tests positive in a house of 10,000 the whole house is killed.

This protocol stems from the notion that disease just jumps on poor unsuspecting individuals and the farmer can do nothing to prevent it.

Another good question to ask the Secretary would be this: "Can our farmers and food system do anything to increase immunity in animals?"

There again, the Secretary only sees immunity as coming from the end of a vaccination needle. That's who takes him to lunch—the vaccination needle people. People like me are such a miniscule part of his world, that we do not even register on his radar. He doesn't know that we can defeat pinkeye with seaweed. He doesn't know that we can defeat poultry respiratory problems with pasturing. My world and his never overlap. The result is that every single solution emanating from the USDA is essentially an endorsement of the industrial paradigm.

Better hygiene, better diets, more exercise, more sunshine, daily salad bars—we know that all of these would bolster the immune system. And yet the powers that be never consider that as a viable alternative because the CAFO is the only viable production model. They really believe, like our Virginia Commissioner of Agriculture Mason Carbaugh in the 1980s wrote in his annual bulletin, "If we went to organic farming, we would just have to decide which half of the world would starve." These officials actually believe this. They make statements like this and go to church and feel exceptionally good inside about protecting the world from lunatics like me.

How did he come up with such a silly notion? Simply. The land grant college researches organic production this way:

· Pick several plots of ground.
· These plots have been used previously in research regarding herbicides, pesticides, and chemical fertilizers. The soil is dead.
· Identify three as the organic plots and three as conventional plots.
· Organic plots receive no soil amendments; conventional receive full complement of fertilizer, pre-emerge weed killer, grubicides, etc.
· Plant hybrid corn—genetically selected to require chemical fertilizer—nutrient vacuum cleaners.
· Leave the plots to their own devices. Organic plots get weedy. So what. Herbicide kills weeds in conventional.
· Harvest corn.
· Measure yield. Organic pitiful; conventional excellent.
· Extrapolate production volume over worldwide corn acreage.
· Conclusion: Organics will kill half the world's population.

Folks, I'm not making this up. This is exactly how the research is conducted. Now can you see why science is not objective? Anyone with a lick of sense understands that chemicalized soils take at least three years to detoxify. The native soil flora and fauna take a long time to recolonize dead soil. Notice, too, that they didn't plant hardy open pollinated corn. And nobody cared about the nutritional quality of the corn; the only thing that mattered was volume. This is Modern Science 101, the foundation of official press releases and government-sponsored research.

Our most recent Virginia Commissioner of Agriculture, J. Carlton Courter, told me at a hearing in Richmond that "raw milk is as dangerous as moonshine." First of all, I'm not sure moonshine is all that dangerous. Our country got along fine without any alcohol regulations until Prohibition, which was a huge mistake. A winery owner recently told me that after Prohibition the only people who knew the alcohol business were the bootleggers. The state government invited these guys to come down to Richmond and help write the laws governing the legal sale and transport of alcohol.

The result was a codified distributor network that to this day stifles our struggling wineries. In Virginia, a wine producer cannot take his wine to town and sell it to a store. He has to go through a licensed distributor. This winery operator said it put the in-state wineries at a price disadvantage to imports from states that do not have these restrictive requirements. Just another example of how the regulations protect an entrenched parasitic entity and stifle innovation.

The British reaction to hoof and mouth, or foot and mouth, is a good example. A few years ago Britain destroyed so many cattle during the outbreak that the stench of the burning animals permeated the entire country and virtually eliminated tourism. I'm told by insiders that if any animal in Virginia gets hoof and mouth all animals will be destroyed within a several-mile radius. That sounds about par for the course.

Sir Albert Howard, British godfather of modern day composting who spent his career masterminding the agriculture experiment station in Indore, India during the 1920s and 1930s, lived through a particularly dreadful hoof and mouth epizootic there. Well documented in his classic <u>An Agricultural Testament</u>, he proved that animals with healthy immune systems did not contract it and those fed what he called artificially fertilized forage were vulnerable. The ones fed his compost-fertilized forage were immune. His conclusion: hoof and mouth only affects nutrient-deficient animals.

Such research, of course, indicts the very pillars of our technologically-advanced artificial food system. The thought of giving up such sacred tenets is just too radical to contemplate. I wonder if all those devastated farmers, and all those burning cows, and all those stinking neighborhoods would trade their horror for an immunologically-enhanced animal agriculture? The tragedy of the whole story was that the answer was discovered by one of their own, more than half a century before.

If it actually succeeds at anything, NAIS will merely destroy thousands of small producers. In that, I predict that it will be successful. Just like slaughter house reform through the HACCP plan destroyed nearly half of the slaughterhouses in the U.S. within 24 months of its implementation. Hazard Analysis and Critical Control Point (HACCP) was instituted in about 2002 with a 3-year phase-in requirement as a response to increasing food-borne pathogens. The previous overhaul was in 1967. Both of these were presented by the Food Safety and Inspection Service (FSIS) as scientifically-based, updated modifications to what purportedly had become antiquated and insufficient inspection requirements. I remember well in 1967 when my Dad and a couple carloads of Augusta County farmers traveled to Richmond to testify against the proposed regulations. They lost, and within two years, our county lost half a dozen of its community abattoirs. In most cases, these were just neighbors who had some equipment—maybe a hog scalder, some knives, a meat saw, some cutting tables, a walk-in cooler—who had a knack for butchering animals.

That's not a skill that everybody has, and these folks were tremendous assets in the neighborhood. I remember going over with Dad to get a hog butchered or a beef done, and other people were always hanging around swapping stories and helping out.

301

The whole atmosphere was social as much as work oriented. Customers walked around the premises and self-inspected. Anyone who was dirty just didn't get patrons. Now, they are all gone, and it's a rural tragedy of monumental consequences.

When the HACCP proposals began a little more than 30 years later, I predicted a bloodbath (no pun intended) among the next level of community abattoirs that had survived the 1967 purge. Industry leaders scoffed at the notion, bowing obediently before the alter of scientific progress. The whole rationale behind HACCP was to let processing facilities write their own plans and police themselves. And as long as the written protocol was approved by the FSIS, then an inspector would not have to hover around as tightly. Promoted as the new science-based approach to replace the "scratch and sniff" inspection, HACCP actually piled mountains of paperwork and infrastructure requirements on existing facilities.

Touted as a privatized industry/government partnership, a truly revolutionary way to insure food safety, it has had exactly the opposite effect and denied millions of consumers the opportunity to purchase meat and poultry from local farmers.

I had a friend with a fairly good sized cow herd who said he wanted to build an abattoir to serve his area. After researching it for a year, he called me completely devastated: "Joel, I need a $500,000 facility to get one pound of ground meat to a neighbor. By the time I put in the handicapped parking, all the bathrooms, concrete, and stainless steel, I just can't do it." And he never did. His entire region is still denied his farm's excellent meat.

Whatever your political persuasion, and whatever your faith in the government, I think we can agree on this: Putting that kind of hurdle between a farmer and his neighbor, and denying the neighbor and her children local, cleaner, more nutritious food, is just plain wrong. NAIS will do exactly what these other alleged reforms have done: Create another barrier to local food systems, and drive more small performers out of business. Such decentralization would do far more to halt disease than a centralized NAIS.

One additional point about NAIS bears mentioning: imagine the potential for market manipulation by having exactly how many animals of every species and their location knowable at a moment's notice in real time. The ability of powerful interests to interfere with natural market conditions would be unprecedented. The many suits and countersuits already in courts articulating the collusion of large packers, for example, illustrates that the current level of data and power is unhealthy. To multiply this potential exponentially is outrageous.

Of course, NAIS advocates say all the numbers will be confidential. But think about data compromises that have already occurred. The Veterans of Foreign Wars had a big one. And many never make the national news. Our bank just called us last week to tell us that a credit card Teresa used at Christmas was compromised by an internal security breach at a large clothing store. The card had to be destroyed and re-issued.

If bench league hackers can penetrate financial security firewalls, does anyone really believe that powerful interests would not be able to penetrate the NAIS database? Imagine the temptation and desire for a large beef packer, for example, to know exactly how many 20-month-old steers are located within 100 miles of the processing facility? It should make any red-blooded American shudder.

The cattleman leader I spoke of earlier said that anyone who is against NAIS apparently is ashamed of his cattle. Anyone willing to stand behind his animals should want to take the number, he said. As if NAIS is the only way for me to endorse my animals.

What a crazy notion. My customers know me. They know my animals. I know my animals. Ultimately, people buy based on the integrity of the seller, whether that's Tyson or Polyface. To say that I'm ashamed of my animals if I refuse to take a government license is the ultimate socialistic notion. Since when did the government license become the final arbiter of quality?

That's why all medical doctors issued a government license are superior to homeopaths and alternative wellness practitioners who don't have a license. That's why homeschoolers whose parents don't have teaching certificates can't learn anything. Give me a break. The tragedy is that this guy probably votes Republican. But when big business stands to gain, the Republicans line up. That's true patriotism. In fact, as the NAIS effort heats up, I'm sure those of us who oppose it will be branded un-American, unpatriotic.

I'm sure these guys would say that the American Indians who didn't want to go onto the reservation were ashamed of their ancestry. After all, if you're proud of your people, go meekly and obediently to where the great White Fathers tell you. Quit being a misfit. We all need to work together here, you on the reservation, me on Wall Street. Let's love each other.

The common theme from the pro-NAIS crowd is that they want to help us farmers. If I've heard one extension agent say this, I'm sure I've heard it a dozen times, "We can't help you if we don't know where you are."

One told me, "If we have an outbreak of a disease, we need to be able to notify you so we can help you. If we don't know where you are, we can't help you."

Although I listened respectfully, what I really wanted to say was: "You people have laughed at our farming practices for 40 years. You have tried to put us out of business. You have printed press releases telling our fellow Virginians that I want to kill half the world. You call me a bioterrorist because my outdoor chickens commiserate with the wildlife. You disparage direct marketing as a noncredible food niche, hobby farming, a joke. You pooh-pooh my natural remedies for pink eye and my composting for fertilizer. And you think I'm coming to you for help? You think I'm going to depend on you to look out for my best interests? Do you have no shame? How about apologizing for four decades of slander, prejudice, and outright lies? How about starting there, you, you, you . . ."

Can you imagine the U.S. Cavalry asking Geronimo to register his whereabouts so they could help him? "Please put on a tracking device. If one of your squaws breaks an ankle, we can't help you if we don't know where you are. If you get short of food, we can't bring you MREs if we can't find you."

Some people may not like the way I connect these dots, but this is how many of us out here in the hinterlands see this issue. The bottom line: If all I knew about NAIS was who was for it, I'd be against it. All you have to do is look at the signatories and you can tell quickly which side truth is on. "Can two walk together except they be agreed" is not only Biblical, it's accurate. Permutations like "birds of a feather flock to together" and "a man is known by the company he keeps" are not bad ways to make a judgment if we're faced with an issue we really don't know much about.

The point is that if I knew nothing about NAIS except the people who are promoting it, I would be opposed to it. That is not foolish. That is sensible. Call it guilt by association. And on some of these issues, it's a way to make a decision fast and not get bogged down in all the rhetoric.

If the industry wants it, let them have it and pay for it. I have no problem with farmers who want to participate in an identification system doing so. That's fine. If their market wants it, and they want to pay for it, and they have faith in it, let them proceed. But the moment the USDA administers it, the whole ball game changes. At that point, government and industry collude. Integrity and freedom lose when that happens.

I resent that the USDA says my cows are part of the "national herd" and that my chickens are part of the "national flock." Whence this nationalistic talk? This erosion of personal responsibility and autonomy is frightening. It threatens the underpinning of our great American experiment. It's like the educator I heard dedicating a new school a few years back say, "Every child belongs to the state." I was shocked. But nobody else was. It just went over their heads like a TV commercial. I almost fell out of my chair.

Where is the outrage? The duplicity of the American public to accept such outrageous anti-freedom verbiage is astounding. I am deeply dismayed by the number of farmers who voluntarily sign up for these invasive programs, fully trusting industrial and government agents. These same farmers, gathered in the morning around coffee cups at the local diner, routinely complain about every government agency you can imagine.

They decry waste. They decry fraud in high places. They decry bureaucracy. And yet when it comes to this, they are like sheep led to slaughter. When the king says to bow, they bow. Fully complicit. Fully zombified. Fully feudal serfified. A peasant class.

Bow to the lords in their castles. Their banking castles. Their government office castles. Surrounded by protective moats of lawyers ready to gobble up any who would dare question the powers that be. The more things change, the more they stay the same. Yesterday it was the promise of chemicals. Today it's NAIS. Tomorrow it will be something else. Just look at who's promoting the agenda, and you can always come down on the right side of the issue.

Chapter 22

Mad Cow

"Your steer had abnormal teeth, so is assumed to be 30 months old, so we will charge you an extra $500 to butcher it," said the secretary of the local federal inspected slaughter house we use. That meant that the steer would cost $1,000 to process rather than the customary $500. She had called me on the phone to tell me.

"Well, we won't make any money on that one," I said to Daniel when he walked into the house as I hung up.

The government's response to the Washington state cow imported from Canada that had bovine spongiform encephalopathy (BSE—or mad cow) in December of 2003 is a perfect study in knee-jerk policy responses. The official USDA word on what causes mad cow is feeding animal parts, and especially herbivore animal parts, to cows. I'm using the term cows here loosely and generically because it's easier to say than bovines. All bovines—heifers, steers, calves, cows, bulls—are susceptible to the disease.

For the record, an entire body of information exists in unconventional research that indicates the disease has nothing to do with feeding ruminants carrion. This research suggests that it is not communicable and coincides with heavy insecticide use (like for fly control) and heavy metal toxicity—especially aluminum. And that is why it is especially prevalent in wild elk herds around old mining sites. I actually don't know which side to believe, but tend to believe this more nonofficial rendition. Are you surprised?

Anyway, for the sake of this discussion, let's assume that the official government-sanctioned cause is true. When did farmers begin feeding dead cows and other carrion to cows? That has certainly not been a traditional feed source for herbivores, which are designed to eat forages. And until very recent times only ate forages.

This new feeding regimen came on the heels of industrialized agriculture. Before that time, neighborhood abattoirs were small enough to dispose of their offal by composting, burying, letting the buzzards feast, or in some cases feeding to omnivores like hogs. My old timer neighbor told of field dressing steers and dumping the paunch into the hog pen. This is a natural order, even though it may sound disgusting to some today. Hogs are omnivores, not herbivores.

As slaughter houses grew in size, the waste stream became bigger and problematic. This created another business opportunity—rendering into byproducts. Finally, huge rendering facilities developed. Ours in the Shenandoah Valley had such a huge grinder that a whole cow could be dropped in at a time. If a cow died, farmers would call the rendering plant and for a fee, a truck would come and pick up the carcass.

After being ground, the material is cooked and separated into different components. Bones can be screened out, steamed, and pulverized into bone meal. This was routinely used as a high calcium supplement for cows, as fertilizer for lawns and gardens, and as feed additives for poultry and hog rations. The precipitate was dehydrated as meat meal and sold as fertilizer and protein-rich feed additives.

After the British mad cow outbreak, countries around the world, including the U.S., began banning animal byproducts in herbivore rations and the entire rendering business collapsed. The little slaughterhouse we use saw a $25,000 annual income from all their offal change to a $25,000 annual expense. Instead of being paid for their offal, they had to pay the rendering company to come and pick it up. For many small plants, this was financially devastating.

On our own farm, we used meat and bone meal in our poultry rations until the mid 1990s when it became impossible to find chicken-free material. Up until 1990, you could purchase meat and bone meal that was species specific—beef, pork, poultry. Feeding carrion to poultry is completely natural. In fact, old timers tell me that one of the first man-sized jobs that farm boys used to have was going out in the winter and shooting or trapping a possum or rabbit for the chicken yard. In the winter, the chickens did not get fresh animal protein from bugs and insects. To compensate in the winter, then, farmers would try to hunt down a small animal, rip open the abdominal cavity with a knife, and throw the carcass in the chicken yard.

This weekly supplement maintained health throughout the winter. It could be likened to some citrus preventing scurvy on the old sailing ships. Even today, on our farm, when we shoot a deer we dump the skeleton in for the chickens to pick off the bits of remaining meat. It's a tonic for them. But it is completely within the natural order of what eats behind what. Birds have always followed herbivores and picked up after them, cleaning carcasses and picking through the manure. Herbivores do not eat dead birds or bird manure.

We began to notice a marked decline in broiler performance heading into the 1990s coinciding with the increased percentage of poultry rendering in meat and bone meal. We looked and looked, and we simply could not find meat and bone meal that was poultry-free. About that time we discovered Fertrell in Pennsylvania and the incredible intestinal photographs from nature's nutritionist extraordinaire Jerry Brunetti. His research proved that the conventional and industrial thinking about protein was completely off base. Minerals are the key to unlocking proteins.

Armed with that information, we ran some side-by-side tests and quickly proved the efficacy of the minerals approach and discontinued using meat and bone meal. As the poultry industry expanded, its waste stream finally became comparable to the bovine waste stream. Poultry byproducts were in everything.

In fact, the confinement houses generated such a concentrated volume of manure that it became toxic. As the industry began desperately looking for some way to get rid of all the manure, the land grant colleges began researching feeding options. Nitrogen in manure and protein in muscle tissue are closely linked.

In feeding trials, the researchers were able to use poultry manure as nearly half of the beef dietary ration and found that cows would eat it. They mixed it with silage, added some molasses, and the cows ate the dead chicken carcasses, manure, sawdust bedding—everything.

Immediately thereafter, of course, in addition to telling farmers to use rendered beef guts in their cattle rations, the USDA began hosting freebie dinners for farmers like me to promote this new chicken manure feeding technique. During the winter months, when farmers attended their seminars, the newspaper contained almost weekly announcements of manure feeding USDA-sponsored workshops. Farmers by the thousands began adopting the practice because it represented cheap protein. In formulating rations, starch like corn is cheap and protein, like soybeans, is expensive. That's the same way it is for humans: potatoes and corn are cheap; meat and dairy are expensive.

At that time we began selling at the local farmers' market and we used the mantra, "Our cows don't eat chicken manure." Customers were appalled. "You mean they're feeding chicken manure to cows?" They would wrinkle up their faces in disgust. I realized then how myopic all of us are. The newspaper and radio farm shows publicized these seminars with big headlines and bold print. The researchers were proud of their findings and they certainly did not attempt in any way to keep their discoveries hidden. It was out there for all to see. Everyone should have seen these headlines about feeding chicken manure to cows.

But they didn't. The rule is that unless it's my world, I don't see. And that was the great lesson for me. I read what touches my world; I listen to what touches my world. Or at least what I think touches my world. That is why I encourage people to read eclectically. And that is why when Dr. Weston Price, whose dietary findings inspired Sally Fallon to start today's godsend, the Weston A. Price Foundation, when asked before he died how his work could be carried on, he replied, "You teach. You teach. You teach."

I did not buy into this new feeding regimen. I knew that cows do not naturally eat carcasses and manure. In fact, shortly after this, our small slaughter house quit buying Shenandoah Valley beef because the owner said he was tired of walking into a cooling room that smelled like chicken manure. The meat smelled like chicken manure. As a result, this new cheap feeding system resulted in the only federally-inspected slaughter house in the area discontinuing its local farm beef purchases. This processor now buys beef from far away in another state and local farmers lost a significant market. Talk about shooting yourself in the foot.

But the USDA researchers continually assured everyone that this new system had no affect on the meat. The meat was scientifically proven to be just like what our ancestors ate. Only now we could produce it faster and cheaper, and that's always a good thing. This is why our philosophy is more important than our science. Our heart trumps our head every time. In the final analysis, we do what we believe, not what scientists say.

Our heart is a filter to what our head will believe. That is why I don't argue with people. Hey, if you think industrial agriculture is the way to go, fine. We come to truth one step at a time; the process can't be hurried. And that is the reason why this book is deeply personal and relies on stories. Ultimately, our heart draws us to our action, not our head. I have never argued anyone into thinking differently. Never. And I'm no slouch as an arguer. I only touch people who either already agree with me or who are at least not opposed to my message.

To a conventional scientist in our western reductionist disconnected fragmented compartmentalized mindset, a cow is just a pile of molecular structure to be manipulated however human cleverness conceives. And any problems we create, our cleverness can overcome. But to me, feeding a cow dead cows, dead chickens, and chicken manure violates the very cowness of the cow. It would be like feeding a human grass and plastic.

Now here we are, a few decades later, and this huge worldwide "OOPS!" is coming off the lips of every official who promoted this type of feeding. The animal byproduct-mad cow link is supposedly undeniable in official circles. Doesn't it seem somewhat disingenuous that the very people—the USDA—who gave us the problem are now promoting themselves to the peasants as the repository of food safety?

Rather than repenting in sackcloth and ashes, these officials don't even skip a beat. They just go right on with business as usual and begin enacting regulatory reform to maintain this safe food mystique. No shame.

One of the first mad cow inspired regulations that came down the pike concerned Specified Risk Materials—nerve-heavy parts of the cow. These included spinal cords and brains. SRMs are now officially excluded from the food system. My question is were we so hungry before mad cow that we needed to eat these things? Really.

Next, and this takes us back to the opening statement of this chapter, the officials determined that mad cow was not really an issue in animals younger than 30 months. So far, it has not been detected in any animals younger than that. Of course, most scientists believe mad cow has a long incubation period and the symptoms just don't show up in young animals. Be that as it may, the 30-month rule is hard and fast.

Animals older than 30 months must be slaughtered separately from animals younger than 30 months. The older animals must have their spinal cord removed and other precautionary procedures, including paperwork, that add significantly to their processing cost. They determine if an animal is 30 months or older by looking at the teeth. Bovine teeth are just like human teeth—they don't all come in at once.

But here's the problem—neither do they come in on a hard and fast schedule. Any two children can have literally years of difference between how their teeth come in. Cows are probably a little more regular, but wide variations exist. At any rate, this hard and fast 30-month rule has literally come to dominate the beef slaughter industry, and it's worse because the age is determined by the teeth. And the particular teeth that come in around this time can come in anywhere from 26 months of age up to 36 months. As the central component of a food safety system, a measuring device that has a 33 percent error rate leaves something to be desired.

This unbending rule becomes a huge problem for grass finished beef producers like us because the genetics that encourage easy grass fattening are the same ones that encourage early pubescence. The genetic link between early fertility, easy finishing, small body phenotype, slick hair, and tender beef is well documented. Farmers like us have been genetically selecting our animals for decades for early pubescence which means these teeth are coming in early. Since we do not finish them on grain, we need a little extra time to get the marbling needed for succulent taste and texture.

The result? Here we are, the only portion of the beef industry that can guarantee that we don't have mad cow because we've never fed them animal byproducts, and yet the system penalizes us for it because our cows get their teeth earlier. In other words, the government response to mad cow discourages the only guaranteed solution. This is by their definition, not mine.

The federal inspector was in the abattoir the other day when I brought some animals in and I asked her about this. "Why can't I sign an affidavit saying I've never fed animal byproducts to these steers? According to your own research, that makes them immune to the disease and I should be exempt from the 30 month prohibition."

"They don't recognize any production difference. It's the teeth."

"But these teeth come in at different times. Why should I be denied that extra four months of growth time? Sometimes I can't get them finished in that amount of time."

"I know they come in at different times, but that's the rule. I'm just doing what they say."

"Well why don't you tell them this is unfair? Why don't you go to bat for farmers like us?"

"I'm just doing my job."

It all comes down to that, doesn't it? The official bureaucrat cop-out. "I'm just doing my job." Or the permutation: "I don't make the rules. I just enforce them." A million ways exist to say the same thing, but it all boils down to blindly following foolishness and feeling no culpability in doing so. For people like us who have to make our living by thinking, these conversations are maddening. These officials get their steady paycheck whether they think or not.

Even with all the hot air blowing out of Washington, farmers in my community, as of the spring of 2007, are still feeding dead chickens and chicken manure to their cows. You can drive right down the road and watch them take a front end loader scoop of silage and dump it in the feed cart, then a load of chicken manure, which of course contains plenty of chicken carcasses. I'm confident the average American consumer purchasing beef out of the supermarket meat case has no clue that out here in the hinterlands cows are eating this stuff.

Creekstone Farms in Missouri lost its Japanese market when mad cow struck. The processor went to their Japanese accounts and secured a promise that buying could resume if every animal was tested. The processor began doing just that, and USDA promptly sued them. The reason? "That would make all the other beef not being tested suspect in the consumer's mind. If one plant checks all of them, then all plants will have to and that's an unfair burden." News flash: After 2 years in court, Creekstone won. The USDA will appeal. Are you surprised?

Of course, mad cow is the single largest impetus behind the animal identification proposal. But wouldn't you think that if an individual business wanted to personally bear the burden of testing each animal for mad cow in order to resume an export stream for American beef, that the USDA would be all for it? That the government would use its power to sue this business for trying to satisfy its customers and create more market opportunities for American beef is unspeakably evil. That's right, evil. And I'm sure many of those officials go to Rotary Club and get their service pins and their Sunday School attendance pins from their churches and think they are protecting all the other processors from being one-upped. What great patriotic Americans, protecting all the other processors from this upstart outfit. How noble and wonderful.

Instead, even though Canada continues to have additional cases, USDA is moving mountains to get the border completely opened again. Irregularities in the Canadian feeding ban abound, and yet they are being given concession after concession after concession. But let some little independent domestic processor try something innovative, and the power of the U.S. federal government comes crashing in on them.

Naively, I suggested to our federal inspected processor that the teeth be checked before the animal goes in the kill gate so that if it doesn't meet the 30-day rule, we could load it back on the trailer and take it to the custom butcher, where so far the 30-day rule is not being enforced. The reason is because under custom, your own animal is being processed. If I raised the animal and fed it the way I wanted to, presumably I should be able to risk eating it. This loophole may close soon.

Problem: I can't reload my animal and bring it back home. I can't go to the federally inspected slaughter house and get my own animal back. I kid you not. Every day I learn about something new that I can't do. And that day I learned about a new one. Once I unload those animals, even though they may not be killed for up to four days, I cannot change my mind and go back and retrieve them. It's illegal. Every single animal that walks into the corral at a federally inspected slaughter house must be killed at that facility—they can't be reloaded to go anywhere else.

Now, on our farm, we religiously check teeth before going to the abbattoir. If they are anywhere close to questionable, we go to the custom butcher with them for our customers who are buying split halves, halves, and wholes. Only clearly defined ones go to the federal facility.

The whole mad cow/national identification system is a study in government ineptitude and intrigue. NAIS was first touted as a food safety issue, then an animal disease issue, and now clearly as an export issue. But I think the whole export issue is a smoke screen since the U.S. is importing nearly five times as much beef as it is exporting. In 2005, the latest year figures are available to me, the U.S. imported 3.6 billion pounds of beef and exported 750 million pounds.

Why don't we just eat it all domestically? Because agriculture has always been the sector expected to lift the U.S. out of its balance-of-trade deficits. In fact, Agriculture Secretary Mike Johanns said, "I believe a fully functional animal tracking system will keep us competitive in international markets, helping us retain and expand our market share."

This from the government that pooh-poohs the European Union for not taking genetically engineered American food. And the same government that castigated Europeans for refusing steroid-laden American beef. When the customer asks for something that disparages the American industrial mindset, the customer is ridiculed for believing old wives' tales. When the customer asks for something that puts the industrial sector in the drivers' seat, that's a noble, scientifically-based request.

The incidents of these kinds of animal diseases are far more numerous in wild herds, yet no national monitoring system has even been suggested for game animals. Brucellosis, Pneumonia, Johnnes, Rift Valley Fever, Creutzfeld-Jakob's—all of these currently reside in and are transmitted by wild animals. This is a known, not a possible occurrence. And yet hunters are not required to submit to inspection.

The bottom line is that the USDA gave us mad cow—according to the official story—and is now doing everything possible to destroy those of us who have the only guaranteed answer. In fact, an answer according to their own official science. They say that mad cow is not transmissible and cannot be acquired by bovines unless they eat animal byproducts. You would think they would try to eliminate animal byproduct feeding and make it easy, indeed encourage, those of us who have been grass-only feeding for decades, to get to the market with our animals. But that would be too reasonable.

However you slice it, mad cow is an industrial food system byproduct. But nothing the USDA envisions to combat it will deal with the industrial food system. Any policy put forth will stifle the answer and further the problem. That's the nature of the beast called the USDA.

Chapter 23

Animal Welfare

"**P**lease write your Congressman and tell them to vote no to the proposal to ban postal shipment of live poultry." The plea comes in email, phone calls, special letters. The entire non-industrial poultry movement depends on the freedom to ship chicks in the mail. But this freedom is constantly being attacked by the animal welfare lobby.

As a Farm Animal Certified Humane producer and a passionate promoter of pigness and cowness, I consider myself a poster boy for animal welfare. And yet I'm amazed at how many times I feel threatened by the animal welfare agenda.

Without a doubt, the animal welfare movement in this country has been stimulated by two things: industrial factory farming and urbanization. Had farm animals never been cramped into CAFOs, the inhumane description would have never seen the light of day. Farmers have been their own worst enemies in this regard. Industrial farmers don't seem to have a clue that for all their platitudes about efficiency and feeding the world, they can never gain the high ground morally for a production model that despicably abuses animals.

Factory farming simply pours gas on the animal welfare flames. And rightly so. But as with all reactions, this one goes overboard too. And therein lies a great danger. Which brings us to the second point: urbanization.

When the only interaction between people and animals is in a pet situation, it jaundices the historical and natural relationship between the two. As someone who has slaughtered literally hundreds of thousands of animals, primarily chickens, I take no delight in the process. By that I mean I understand that my predacious existence does not allow abuse.

Unlike a cat, which relishes the chance to play with a mouse for a long time before finally killing and eating it, I do not scratch, beat, and paw animals before killing them. We honor and respect them both in life and in death. And just for the record, humans are not animals. Animals don't sin. Animals don't have souls.

This notion that modern humankind has evolved beyond killing animals is simply the result of too many people being totally disconnected from life. Anyone connected to life understands the cycles of life, which include death, decay, and regeneration. We've raised a generation on Bambi and Thumper rather than Thanksgiving hog killin' and the Christmas goose. Those who say they've achieved a spiritual Nirvana by being at one with the animals are only showing just how disconnected they are from real life cycles.

To be sure, I have no problem with vegans or vegetarians. I have no problem with animal worshippers—the ones who say a person is a cat is a fly is a grasshopper. The problem comes when they try to use the political process to outlaw meat consumption. Interestingly, these folks vilify the religious right for trying to impose their ideas on others, but have no problem when the shoe is on the other foot.

This became quite apparent to me in the early 1990s when I was asked by the Humane Society of the U.S. to help write the humane standards for a wonderful book they put together called the The Humane Consumer and Producer Guide. In my mind, this is still one of the best national directories ever compiled to connect humane farmers with people who want to buy their food from these kinds of farmers. Anyway, in the standards, it was considered inhumane to abort a fetus from a heifer in the third trimester of the pregnancy.

If a heifer is not going to be used for breeding, farmers will sometimes abort the calves in order for the heifer to gain faster as a beef animal. Early in the pregnancy, drugs can be used. Often heifers are spayed. But if a heifer enters a feedlot already heavy with calf, farmers will induce abortions in order to simplify their operation and put the calories on backfat rather than into milk and a baby calf. Amazingly, the people who are so concerned about abortions in the third trimester of a bovine pregnancy tend to support that action in humans. Isn't that incredible?

It's as inconsistent as the pro-lifers eating disrespected, factory farmed meat out of Costco. I agree with Matthew Scully, author of Dominion: The people who should be most concerned about respecting and honoring animals are the members of the religious right. Instead these folks defend the right to abuse animals, to disrespect their chickeness and pigness. And they even applaud their own ability to find the cheapest food. I wonder if they think the best church comes from hiring the cheapest pastor.

Invariably, when animal rights advocates come to the farm and we begin talking about things like castration or embryo abortion, they always assume that I'm a fellow rabid human baby abortionist. As a farmer who has helped many cows deliver calves, the moment of ecstasy is when you reach in and the calf pulls away. That's when you know the calf is alive. And no farmer ever looks at his assistant and says, "Oh, good, this fetal mass is moving."

Rather, we exclaim, "Oh boy! It's alive! Let's get this little guy out of here." And if it's alive then, was it alive yesterday? How about the day before that? And the day before that? If any animal welfare groups want my respect, they will have to come out passionately in favor of a human pro-life position. In my soul, I cannot see how a person wanting desperately to save a tree or save a baby whale has no remorse at snuffing out a wiggling, very much alive human baby in what should be its safest environment, a mother's womb. The rise of the abortion movement coincided perfectly with our culture's disconnection to the land.

The wonder of life, the mystery and majesty of chicks hatching and pigs farrowing creates a deep appreciation for new life. Even the satisfaction of seeing cows get bred—knowing the value of that developing calf—makes farmers want to protect these developing babies. We farmers know they are the future. They are our survival. We do everything possible to bring those babies to term. People who don't see that routinely, who don't experience that, can easily lose that sense of awe. And when babies no longer instill awe, we've not become a higher developed society, we've become crass and harsh.

I don't know how many people have said to me, "How can you butcher those animals? I just don't think I could do it. They're so cute." I could just as easily turn the question around: "How could you live in a townhouse divorced from fields and woods and vibrant life? I just don't think I could do it. It's so sterile and dead."

A lot of this is just in the way we've been brought up, what we've experienced. I've grown up on the farm, battling possums and raccoons. And rats. Oh, I hate rats. When you find a hundred half-dead chicks stuffed down a hole under the brooder house, you develop a keen distaste for rats.

We had a guy come to one of our seminars sporting a PETA bumper sticker. That stands for People for the Ethical Treatment of Animals. Out here in farmland, we call it People Eating Tasty Animals. My favorite is this rendition: PETA—Indian word for poor hunter. Anyway, this fellow decided that if he couldn't kill it, he shouldn't eat it. Because every time we eat meat, we are vicariously taking an animal's life. He had been a vegetarian for several years.

The first morning of the seminar, we gave him a knife and let him kill some chickens. He appreciated our honoring them in life, and our honoring them in death. It was quite an epiphany for him, and he ate chicken the next night. Can you imagine a Tyson slaughter plant allowing this guy to come in and kill chickens? They have all sorts of no trespassing and security signs posted.

This is part of the problem with the food system. We have made it unfriendly to people, to the extent that people can't be connected to it even if they wanted to. This inherently breeds disconnect, misunderstanding, and mistrust.

On our farm, we've changed the name of our pastured shelters from pens to shelters, partly as a response to this animal welfare movement. Pens sound like a shortened version of penitentiary. It has all sorts of negative connotations. In farm country, pens do not carry that kind of negativity. When we put animals in a pen, it's usually because they are receiving special care. If they aren't in a pen, they are just out there on the range so to speak, fighting the elements and surviving with minimal care. But in a pen, that's where they get special attention.

For urbanites, however, pen holds an entirely different meaning. They think it smacks of confinement, being enslaved, penned up and not free to move. In fact, some animal welfare folks visited us and castigated me for having the broilers out in these pens. They thought it was awful. I assured them that if the birds were not in a pen, they would be destroyed by predators and weather. And if one happens to get out, all it does is run around and around trying to get back in. They instinctively know that the pen means safety. And they don't want to be away from their buddies.

As we began realizing our language liability, we changed the word to shelter. Shelter sounds more like nurturing and care. Of course, it's a lot harder to say and we had to work hard at going to all this extra speech effort, but it has paid big dividends. We haven't had any complaints for some time now.

Which brings me to the main point I want to make in this chapter. All of us suffer from the weakness of jumping to conclusions. Becoming acquainted with a topic a little out of our area of expertise is downright difficult. Just plain hard work. I've been pushed to understand ex-vegetarians, for example. I had a customer at one of our metropolitan buying drops pull me aside and discreetly ask, "How do you make a hamburger?"

I said: "You're kidding."

"No. My husband and I have been vegetarians for about a decade until we found you, and now he wants a hamburger and I don't know how to make one."

Believe me, folks, I am not making this up. I confess to having to control myself when these things happen. I need to walk in that person's shoes. Ignorance is not wrong. All of us have our blind sides, our inconsistencies, and our ignorance. And too many folks ascribe to animals human characteristics because they've never been around animals.

For example, plenty of people have been turned in to animal control officers for allowing their animals to be outside in the winter, especially horses and cows. These animals grow exceptionally long winter coats as cold weather approaches. Actually, they are often much more content out in the open air than they are cooped up in a dank, dark, confining barn.

I had a friend who received a visit from an animal control officer based on a complaint from a neighbor that his horses were freezing. What the neighbor saw was the steam rising off the horses in the morning sunlight. This same neighbor had a pond on which several ducks were happily swimming around.

"Do those ducks look cold to you?" he asked her.

"Well, no, they're perfectly content."

"Then why do you think my horses are cold?"

"Well, if I were out there like those horses, I think I'd be cold."

"Well, if I were out on that pond like those ducks right now, I'd be cold," he responded, good naturedly.

Suddenly she understood his point, and apologized for sending the animal control officer over. As a matter of fact, if she was truly neighborly, she should have walked over and talked to the horse owner before calling the bureaucrat. Wouldn't that have been the charitable thing to do? I sincerely hope that anyone reading this who would ever invoke the power of bureaucrats on a neighbor, would at least first go to the neighbor in person and express the concern. This behind-your-back calling a government agent on a neighbor is unconscionable among civilized people. Goodness, it's not acceptable among uncivilized people either.

Back to the chicks in the mail. This is a similar situation. The well-meaning animal welfare advocates have this idea that babies shouldn't be transported in a box without food and water. Somehow this is inhumane. Poppycock. Chicks do fine for 72 hours without food and water. They aren't human babies. They are completely different. A calf can't and a baby pig, baby rabbit, or foal can't. But birds are different. They aren't mammals, and they can handle it just fine because they don't nurse. Nursing babies need immediate attention. Birds don't.

This effort among animal welfarists to shut down the transportation of chicks and poults is currently consuming enormous amounts of time and siphoning off energy in emotional fear from the small-scale poultry community. Would the animal rights folks prefer to shut down the pastured poultry movement so that only vertically integrated operations that ship enough chicks to justify their own fleet of trucks would be the only poultry available? Are factory farm transported chicks superior to mail-order chicks for pastured poultry operations? That's a rhetorical question, by the way. The answer is obvious.

As a farm that receives thousands and thousands of chicks in the mail, I have my horror stories. In fact, one of our mail carriers said after a disaster last year, "I think shipment of chicks should be outlawed." Here is what happened. We had a heat wave. When that happens, the hatcheries pack the chicks very loose in the box so they can get plenty of ventilation. The chicks were fine when they arrived at the distribution center in Richmond, Virginia.

But in Richmond, for the short leg to our post office, the workers put the chicks on a truck in the afternoon and it sat on the paved parking lot for 5 hours before coming to our post office. About 2,000 chicks suffocated. They all arrived dead. Of course, when animal welfarists hear this, they go apoplectic. Rather than being angry with the buffoons at the post office that should have known better than to close the door on that truck, they want to criminalize the whole shebang. What is criminal is postal worker negligence.

We've had postal workers, when they've found an incident like this, rip open the boxes and put a cup of water in each quadrant to help the chicks stay alive. Trust me, I've seen wonderful postal workers and incompetent ones. One time we got chicks in a cold snap and at one of the distribution centers, the workers just set the chicks out on the dock in the freezing cold. Those chicks froze to death. Was that the system's fault? No, it was the fault of an unthinking nincompoop.

Another favorite of mine is farrowing crates for sows. Again, the industry puts sows in crates so confining that the hogs can never turn around. Ever. Their whole life. It's tragic. But these crates do help pig survivability at farrowing because they keep the sow from being able to accidentally trample her babies. Remember, a sow may have up to 20 in a litter.

We buy pigs from local producers who let their sows run free. It's a great life. Right when the sow is ready to pig, the farmer puts her in a gestation stall that confines her for a couple of weeks. This enables the piglets to get up and going before the sow can move around enough to step on them. Certainly some farmers don't use gestation stalls, and I am not trying to defend stalls completely. But the animal welfare movement wants to outlaw them entirely. I humbly suggest that confinement for a few days to increase baby pig livability is worth it. It's not a zero sum game. Many things have tradeoffs. And rather than demonizing the crate itself, perhaps the animal welfare movement would have more credibility if it demonized the confinement when sows aren't farrowing.

The whole farrowing crate issue is complex, and I've met many producers who do not use them. Some simply take a Darwinian approach and let happen whatever will happen. Over time, hardy genetics win out and they gradually build a herd of sows with better maternal instincts: i.e. they don't trample their babies. Others have modified crates so that the sow can exit and enter at will. This hybrid idea protects the babies but also allows the sow to get the exercise and fresh air she needs. Recently I've seen tremendous innovation and movement in this area.

But just for the record, the best piggies I buy are from a fellow who uses crates—only for a couple of weeks. Otherwise, the sows have a better life than the sows of any other producer from whom we buy pigs. My whole point is that these issue are not always just cut and dried like many advocacy organizations try to paint them. For some reason, we humans have a penchant for causes. We love to be crusaders about something. I think it's in our makeup. Unfortunately, too many of us become crusaders for causes that strain at truth.

Another blind spot in the animal welfare community is horse slaughter. I have been deeply chagrined with the effort to prohibit horse slaughter for human consumption in the U.S. I expect that by the time this book comes out, the final abattoir will be out of business. What legislation couldn't do, the judiciary has done. And it's a shame. I couldn't disagree more with my friends in the animal welfare movement over this issue. Although horse meat is not a staple of the American diet, it is consumed, with relish, in other cultures.

To deny farmers the extra value created by a vibrant horse meat sales option just because I don't like the idea of eating the Black Stallion is myopic to the extreme. Talk about the religious right. Give me a break. Since when did horses become sacred over llamas or cows or pigs? I guess since Flicka. A slaughtered horse is a slaughtered horse. Just because the meat goes to dog and cat food, does that make the killing act more noble?

Horses get old and stiff and crotchety. Allowing a useful market when they no longer can stay healthy is only reasonable. To what animal will we ascribe this non-human food status next? Squid? Lobster? And this is what concerns me. When I read the arguments these folks are putting out, it is clear to me that their real agenda is to make all animal slaughter illegal. All of it.

To do that, philosophically, a person must equate animals with humans. And that is an untenable position. Let me explain why from an ecologist's point of view. Tillage is generally destructive to soil. Historically, all sustainable tillage schemes are on a 5-7 year rotation, in which the land is tilled only about 2 years out of 7. The in between years of grass rebuild the soil.

Grass is nature's most efficient soil builder. It's also the most efficient carbon sequestration mechanism. Much better even than forest. Planting annual crops year after year requires large off-field inputs in both organic and conventional chemical systems. In all sustainable systems, tillage only 2 out of 7 years is all the soil can stand without hefty imported amendments. From a land healing and atmosphere cleaning perspective, nothing is as efficacious as grass. Grass value only increases on marginal lands. Millions of acres on the planet are not suitable for crop production, but they grow wonderful forages.

In fact, now we know that today's dense eastern forests did not exist before Europeans came to the continent. The Indians maintained savannahs by lighting routine fires to beat back encroaching trees. This manipulation encouraged more grass to grow stimulating herbivore populations. Manure is magic. Always has been; always will be. Even with all we know about soil fertility, we still don't know what the X factor is in manure that makes it better than the artificially reconstituted elements found in manure.

Without perennial meadows and grasses, we would have a more eroding landscape and a dirtier atmosphere. How do grasslands stay healthy? They regenerate and proliferate through routine mowing. That mowing is most efficiently performed by animals. Certainly some folks would say that animals on the landscape do not necessitate carnivorous humans. And while that may be true, part of the human responsibility is to steward the landscape to make it capture more solar energy, to sequester more atmospheric carbon, and to make it more productive than it would be if left to its natural devices.

Grass-based meat is a whole different nutritional item than grain-based meat. The recent discoveries regarding the B vitamins, conjugated linoleic acid, and the polyunsaturated fats, including the omega 3:omega 6 ratio, are proving that all of the alleged human health problems associated with meat consumption are a result of artificially producing that meat. We've had numerous customers who return to meat after a decade of vegetarianism destroys their health. I suggest skeptics contact the

Weston A. Price Foundation for corroboration, or log onto <eatwild.com> for cutting edge nutritional findings.

Whatever is wrong with eating meat and poultry is a result of producing it in factory farms and feedlots. Whatever environmental degradation, human health problems, or animal welfare issues impugn meat and poultry consumption can be rectified and turned into positives with a fundamentally different production style.

Beautifully, this fundamentally different production style would result in the 70 percent of North America's tilled farmland being converted to perennial grasslands. Only 30 percent of the grain acreage is for people, pigs, and poultry. If we really want to heal the land, atmosphere, and our bodies on a massive scale—not to mention getting the petroleum out of agriculture—the fastest way to accomplish that is to increase demand for 100 percent grass finished beef and milk in this country. That is far more healing than anything else.

Finally, I actually have more respect for true vegans than I do vegetarians if the issue is animals equal humans. Vegetarians who eat eggs and dairy but refuse to eat meat because killing animals is wrong have no understanding of animal life cycles. Where do they think eggs and cheese come from? They certainly don't come from geriatric livestock. They come from productive, virile, breeding age animals. And as those animals age, they must be culled from the herd before they become unproductive.

The symbiotic relationship between grasslands, ecology, and herds of herbivores is a natural principle. It has been functioning for millennia. For anyone to suggest that eliminating these relationships could be normal is to not recognize historical principles. When we begin looking at nutrient density, nothing beats meat, dairy, and poultry. Extricating animals from the landscape is not healthy for anyone or anything.

Certainly some people thrive on a vegetarian diet. I don't know anyone who thrives on a vegan diet. I've met many folks who are on a vegan diet, but I've never met a healthy one. Many times the diet works great as a cleansing or detoxifying regimen, but just because something works great as a temporary curative doesn't mean that continuing it is better. Antibiotics that knock out infection are great, but continuing to take them beyond the point of cure isn't healthy. Virtually all of the supposed animal-protein-induced toxicities are the result of factory farming. Grass based changes everything.

Anyone wanting to invoke religious reasons for not eating animals should realize that Jesus certainly ate meat. So did Mohammed. So did Indians. I realize that in the Garden of Eden the lion didn't eat the lamb, but we haven't been in Eden for a long time. And it won't return because of our imposition; it will return only because outside spiritual forces intervene in our world.

I am all about caring for animals. On our farm, we do chores before we eat breakfast. Always. I would much rather deny myself rather than my animals. But that doesn't mean I worship them, or that they are human.

A chef was out one time to see the pigs and we walked up to the pig pasture. He had never seen live pigs before. His only acquaintance with pigs was pork. These pigs were scratching on trees, rooting in the dirt, lounging under bushes, nibbling at weeds and grass. He stood quietly for awhile, mesmerized by the theater before him and the actors enjoying their parts. Finally he said, "I don't know anything about pigs. But I think if I were a pig, this is the way I'd like to live."

To me, that said it all. And that is the attitude we take toward the animals. I don't pay much attention to the folks who think children are dogs are rats are crickets. And I wish politicians wouldn't either. If our culture continues to destroy direct farmer-consumer local food commerce, people will continue to become more and more unreasonable in their thinking. Supposing themselves to become wise, they've become fools. And fools often pass laws.

That is one reason why we invite and encourage people to come out to our farm to visit, to touch, to see, to smell. Real husbandry on a real place in a real time helps to punch through the academic and theoretical disconnects pontificated by the radical animal welfare elite. I am native. That should not be illegal.

Chapter 24

Options

All of you diehard readers who have stuck with me this long deserve an upbeat close. We've sure been around the horn, and I hope by now even the skeptics can see, through the personal stories of one little farmer, why I feel like I've just gotten sent to the principal's office every time I receive a letter in the mail from the USDA. My list of illegal activities would fill a book—oh, yeah, it did. So where do we go from here? Where's the hopeful conclusion?

Many farmers are using the pet food loophole. Labeling everything as pet food and "not for human consumption" has worked in some areas. These are beginning to close and I predict that trend will continue. But for now, some states have lax pet food laws. By calling everything we sell pet food many farmers are getting around the regulations.

Other states, however, aren't as easy. Some require full-blown nutritional information on pet food—identical to that required for human food. Often, though, these don't kick in until the 9,999th container. My advice is to just go do it. It's always easier to ask forgiveness than permission. And if it goes to 10,000, who is counting? I'm not averse to pushing the envelope.

In dairy, the cow share program continues to gain momentum. This is an arrangement where the customer owns a piece of the cow and pays a boarding fee. The ownership guarantees milk and dairy products from that animal. As these have proliferated, the bureaucrats have been right there to tighten down the screws.

For example, when these first started, a dairy could sell shares that entitled the owner to say, two gallons of milk a week. The farmer would milk the cows and, like normal, hold the milk in a bulk tank until the customers came to get their milk or he delivered it. Now the milk police are demanding that the milk not be comingled. In other words, if I have a deed of ownership that entitles me to the proceeds from my property—Bossy Number 53—then milk from Bossy Number 54 is illegal. That has to go to the owner(s) of Bossy Number 54.

This new stipulation, of course, creates a real hardship for the dairy trying to get customers some milk because it all has to be kept separate out of the cow. That segregation is not too difficult to do if you're milking one or two cows, but when the numbers go to ten or more, it becomes quite difficult. While such a stipulation makes technically legal sense, it is clearly an attempt to stifle the cow-share loophole. Some states are blatantly legislating against cow sharing, calling it a sales charade.

Of course, many states allow the sale of raw milk. The cow share arrangement is only needed in the states that have such a hostile view toward raw milk that they outlaw it outright. The people resorting to this technique, then, are inherently implementing it in enemy territory. The reaction by these states, therefore, is certainly no surprise. Some court cases are already beginning on this technique, and no doubt will continue for some time. Whatever the outcome of the court cases, this technique will come under closer and closer scrutiny.

The problem with cow sharing is that it is inflexible. What if I'm going on vacation and I don't need my milk for a week? The milk cannot legally go to someone else. What if I have a birthday party and I need an extra gallon to make ice cream? I can't just buy it, because someone else owned that gallon. The point is that it is inflexible for both buyer and seller, a logistical nightmare. If raw milk and dairy are as dangerous as prohibition states declare, it should be illegal to drink even from your own cow. It certainly ought to be illegal to feed to your children—child endangerment, maybe? I mean, if it is really as "dangerous as moonshine," wouldn't the Department of Social Services take the child out of a home in which the parents were giving their children white lightning?

Such a notion is the obvious extension of the hazardous material argument. It applies to everything uninspected. And this is why this book should be read by every consumer advocate in the Ralph Nader camp and every patriotic flag-waving American in the Rush Limbaugh Advanced Institute for Conservative Studies. The one group sees all the solutions in legislation, kind of salvation by legislation. The other group see nothing wrong with regulated corporatism, which really destroys the little and encourages the big.

And this is why I take such a militant view toward government involvement, because the same philosophical justification for intervening in raw milk consumption also gives equivalent justification for intervening in uninspected chicken or homemade cookies. In Virginia, nearly all the heritage apple butter festivals have gone by the wayside as nonprofit organizations get letters from their insurance underwriters: "We can no longer protect you in activities not inspected by the Food Safety and Inspection Service." No, it's not a conspiracy. It's a fraternity of ideas.

It's simply a ubiquitous mentality that responsibility for actions are not individual, but emanate from somewhere out there, out in the morass of government offices and paper pushers. This is why I'm in favor of legalized drugs and alcohol. The same thinking that assumes it's okay for government to keep me from smoking dope—to protect me against myself—also justifies the government to regulate my use of Vitamin C or homeopathy. My fundamentalist Christian friends go apoplectic when I say such things, but I would rather a few people blow their brains with cocaine than that my uncle be denied an unconventional medical treatment of his choice.

As soon as the freedom for me to choose one thing I ingest becomes a government issue, then that oversight can consistently be extended to any and all of my ingestion choices. If we can't own our own bodies, then what can we own? When I hear people say, "We need a law" to correct some perceived life risk or supposed societal avarice, I cringe. Just like technology can be used for good or evil, the political process is the same way. And my experience requires me to be dubious whenever the power of government is invoked as a cure.

This brings us to the crux of the whole food safety issue. Aren't you glad you stayed with me this long? Here it is: the political rationale for food safety ultimately rests in the notion that we are wards of the state. Not a free people. And in today's governmentally-invasive climate, micro-managing food freedom can be sold to the American people as a medical insurance plan.

Having talked to many food safety bureaucrats in both Richmond and Washington, I can assure you that they believe they are keeping millions of Americans out of the hospitals by denying them home-processed and on-farm processed food. They think raw milk would fill up the hospitals and the morgues. They think unwashed eggs play Russian roulette with the nation's health.

I've been called a bioterrorist because my pastured chickens commiserate with Red Winged Blackbirds and sparrows, who then take my chickens' diseases to the science-based, environmentally-clean factory chicken houses. Yes, those houses that stink up the entire neighborhood and pollute the groundwater. Yes, those houses that can only be entered wearing dust masks and moon suits.

Ultimately, the government's involvement in medical care creates the justification to penetrate personal liberties with regulations that codify exactly what the wards of the state may or may not eat. Unlike many of my friends who just see evil emanating from the hearts of these bureaucrats, I do not. I see honest churchgoing Rotarians who truly believe their life's endeavor to prohibit food like mine from gracing America's table

335

is the only thing standing between our nation's alleged good health and an epidemic nightmare. God bless 'em.

A government that can control alcohol, cigarettes, gambling, and drugs can and will ultimately control everything we can ingest, all in the name of protecting us from ourselves. Who decides what is damaging? Who decides what is safe? Like one of my customers said to the regulators in Richmond, "I've looked at the stuff in the supermarket, and I've decided most of what's in there isn't safe." You can drink 20 Cokes a day, but be careful about that homemade pound cake—it will surely get you.

Now we're going after transfats. What's next? Homemade jelly? Too late, it's already been demonized. As governments become afraid, they begin criminalizing things. Only insecure despots have to look over their shoulder all the time. I believe this increasing host of regulations is symptomatic of a culture on the run. We're running from people we've wronged. We're running from people we've alienated. We're running from increasing distrust and hostility among the populace. We're running from disenfranchised small business owners. We're running from dehumanized and decultured globalism. We're running from cheap food, obesity, and food borne pathogens.

As we run, we retaliate with law after law after law. We think we can save things from the top down, but the top down doesn't ever save anything. The only safety comes in our communities, our homes, our families, from the bottom up. And these institutions must be free to experiment, to innovate. I confess that I do not hold optimism for our culture as a whole. Indeed, we are destroying agrarianism and seeing the same political graft and moral debauchery that brought the Roman culture to ruins. I am, however, optimistic about the power of one. The strength of you and me and others, touching our spheres of influence, taking individual responsibility.

I will gladly put the society cost numbers of those who take their own responsibility seriously against those who take a victimized mentality and expect government to take care of them. Are we richer or poorer because we let the Amish opt out of social security? Are we richer or poorer because we let home schoolers finally proliferate and opt out of government schools?

As a culture, we have indeed done some right things. And they are shining examples of what true individualism can bring to a society. Being different is not always threatening. Sometimes it's liberating and rich. Are we richer or poorer because so far we haven't clamped down on eBay? Lots of times the greatest breakthrough come from folks who break with conventional paradigms. How a nation treats its lunatic fringe determines its level of freedom or tyranny. What we choose to define as threatening actually determines what solutions we allow.

When we define local food, or uninspected neighborhood food, or pastured poultry, or compost, or Vitamin C therapy, or acupuncture, as a threat to the culture, we deny ourselves the solutions that other belief systems may bring to the table. That does us all a grave injustice, and denies our children a world filled with the innovations that the fringe inherently creates.

This next is one of the most poignant stories of my life, and one that helps put all this in context. On one of our family's forays to Historic Williamsburg, we happened to attend when George Wythe was there. Yes, this is a time warp.

Costumed interpreters take on a historic figure and perform around the world. Many folks are familiar with Hal Holbrook as Mark Twain. Okay, George Wythe received guests in his house for a couple of hours while we were there, and I went to see him. Wythe was Thomas Jefferson's law professor at the College of William and Mary. This man was so much in character, that a modern word like "escalator" would not compute in his mind. He literally transported visitors to his time and place. As one who loves drama, this to me was the greatest way to study history.

When I had my turn to dialogue, I asked him about the Native Americans—the Indians, and his view toward them.

"They are just barbarians."

"Why?"

"They aren't civilized."

"But they have a language to describe every human emotion; they have government; they have treaties and commerce..."

"But they don't have the indicators of civilization. They don't drive stagecoaches; they don't powder their wigs—they don't even have wigs; they don't have cobblestone streets . . ."

"But they have these neat dug-out canoes and they use the rivers as roads. Besides, why should cobblestone streets and powdered wigs be the defining characteristics of civilization?" I pressed.

We went on in this vein for quite some time. He clearly was getting a kick out of my playing along and enjoyed the depth that I was pushing the conversation. Finally, somewhat exasperated, he threw up his hands and said, "Look, I may be wrong, but they are just barbarians. Just barbarians. That's the way I see it."

We left that topic and went on to taxes, in which I told him I came from a country that took nearly 50 percent of its citizen's earnings in taxes, to which he replied, meditatively stroking his chin: "Hmm, that must be a far country. I have never heard of such a thing. I would think it would be time for a revolution."

Indeed.

Back to the barbarians. That was the overriding notion throughout our nation's early European settlement. Now I don't suggest that my ancestors were wrong to come here, but I do think we White Anglo-Saxon Protestants were dishonorable in the extreme the way we annihilated the Indians. Instead of learning what we could from them, we summarily destroyed them. And that is why I feel like Sitting Bull. The bluecoats are still coming—they wear blue pin striped suits and sit at big oval tables in Washington and state capitals.

And they have decreed that people like me who think the cowness of the cow matters, and that the chickeness of the chicken matters, who think these critters should enjoy a native diet and a humane existence—we are barbarians. We represent an unscientific scourge on the landscape. That if you have the freedom to choose your own food and your own farmer, the freedom to opt out of government-sanctioned food, that you are jeopardizing the health and welfare of our great Republic, so help me God. Drum roll, please.

With these laws already on the books and those coming through NAIS or mad cow or whatever else strikes their fancy, these government agents whip the American people into a frenzy of fear so that as a culture, we can stomach rounding farmers like me up one by one from our sacred soils, our loving lands, and putting us on reservations. In such a state, your children and my children will only be able to taste grandmother's cookies in isolated museum-type environments, closely controlled by the guardians of our food system.

Listen to my passionate cry: I am a native American. I represent the best of our land stewardship, the best of our food nutrition, the integrity of our food culture. In the continuum of history when the last direct farm-to-plate transaction occurs, farmers like me and food choice enjoyed throughout history will be relegated to dusty archives. Had we honored our treaties with the Indians, respected their personhood, honored their right to be different, would we have a richer country? Are we not poorer for having taken, plundered, lied, and murdered in the name of creating safety for pioneers?

Would the American experiment have failed if we had honored our treaties? If we owned only half the area currently occupied by the United States, for example, would our experiment have been impossible? Would such respect and honor have translated to other people groups today?

Ultimately, the ability to choose what I will feed my 3 trillion intestinal flora and fauna is the final freedom. Who cares how much gas is in the car when I have no food choice? What good is religious liberty when I may attend church having only eaten irradiated, genetically prostituted, amalgamated, extruded, reconstituted, flavor-added, MSG-laden, nitrate-stabilized Archer Daniels Midland pseudo-food?

What good is preserving social security when my subsistence comes only from a government-sanctioned package? And what good is a house and roof over my head if the dinner table only contains cardboard-tasting, devitalized, outsourced slime? If we do not preserve the freedom to opt out of government-sanctioned fare, we will soon lose the last of our other freedoms.

Make no mistake, some states are already talking about micro-chipping vegetables. The California *E.coli* spinach outbreak in 2006 already has officials and many consumer protectionists advocating for vegetable laws similar to meat, poultry, and dairy. From the stories in this book, I hope it's obvious beyond elaboration that such a plan will be the death-knell to farmers' markets, Community Supported Agriculture, and the entire alternative food movement.

We've already lost raw apple cider. Cider has been a fall ritual in America from time immemorial, but now it's practically gone. And now I see California planning to mandate pasteurization of almonds, without a required label. Folks, it's coming. Throughout my experiences with the bureaucrats, I have watched the produce folks indulge my stories but always pass them off as inapplicable to fruits, nuts, and veggies. Get ready. Pennsylvania has plans to microchip tomatoes. Common folks like you and I can't imagine the machinations taking place in the halls of our bureaucratic office buildings. But just because we can't imagine it doesn't mean the assault stops. It continues and continues

The only reason organic and alternative fruits, nuts, and vegetables have enjoyed the success they have is because they have not been heavily regulated. Put them on par with animal proteins, and they will come to a screeching halt. The proliferation of these italicized pathogens is directly caused by industrial agriculture, including the grain feeding of cattle which acidulates and mutates otherwise benign bacteria. The answer for all these ills is a local, size appropriate food system, including production, processing, and transportation. And that is only possible on a credible scale if government will get out of the way and let us be natives.

340

Finally, a couple of good stories. One comes from a hog farm in Georgia during the hard times of World War II. A farmer there had a thriving business, selling 100 hogs a year to folks in the community. The government said he could only sell 10 hogs one year because the rest needed to go to the war effort.

This farmer happened to be an avid raccoon hunter. He had an excellent coon dog who lazed around the house all day but was a terror to the bandit-faced critters at night. During that year, then, when people would come to the farm to buy a hog, he would tell them he couldn't sell them one. They would spy that coon dog lounging by the back porch and ask the farmer if they could buy the dog instead.

"Oh, that dog's not for sale. He's really good."

"But oh, I have a hankerin' to do some coon hunting. I really want to buy him. What would you take for him?"

"A hundred dollars, and not a cent less," the farmer replied.

"Oh, bless you, bless you. I'll take him." The customer would hand over $100 and the farmer would whistle to the dog to come over. The dog would amble over and the customer would open the car door and the dog would jump in.

"Now, let's see," said the farmer, squinting, pushing his hat askew and scratching his head, "I do believe you're gonna need some food for that dog, and he really loves pork. That's what me and the missus have been feedin.' How 'bout I throw in some so he won't go hungry?"

"That'd suit me fine," sir, said the customer, winking. The farmer would disappear to the freezer and bring some boxes of packaged pork out and put it in the back seat.

"There you go. That should be enough for awhile. "

They bade their farewells and the customer would drive down the lane. At the end of the lane, he would open the door and the dog would jump out and go tearing back up the lane to the back porch, where he expeditiously returned to his interrupted nap . . . until the next pork customer showed up. That farmer sold that dog 100 times that summer.

And finally, a lady in Wisconsin was trying to sell cheese into a Minneapolis Farmers' Market. The food police busted her for not being a licensed cheese operation. The pet food laws were difficult. Finally, she hit on a plan and called the Division of Game and Inland Fisheries: "What is required if I want to sell fish food?"

"We don't really have any regulations on fish food, or fish bait, ma'am," answered the clueless bureaucrat.

"Well, if you did, what would they be?"

"I suppose just that it be edible."

"Thank you, sir. You've been most helpful."

The next week she began taking Fishbait Swiss, Fishbait Cheddar, and Fishbait Colby. Nobody could touch her. And as far as I know, she's doing just fine.

Several farmers in Virginia are now giving their food away. Just taking donations. Of course, gifts are not taxable. End of tax returns, end of social insecurity, end of income tax. Their patrons love it. People love to beat the system. It's a power rush, a real adrenaline high. And these farmers are making far more money now than they did when they were selling things.

Which just goes to show, dear folks, that all of us need to keep our chins up and keep on keeping on. I hope these stories from my heart to yours have taught, entertained, and stirred you to never take dinner for granted. We live in a wonderful country, a wonderful world, full of opportunities and discoveries. Let's use all of our gifts and talents for righteousness and freedom.

If I don't have the right to choose what to feed my 3 trillion bacteria in my intestines, then what other rights could possibly be more important? This is such a fundamental notion that to even say it seems silly. And yet, in our modern post-freedom America, many, many people do not believe such a choice should exist.

And that is why you and I must make individual decisions, every day, to patronize the food system that recognizes this most basic human right—the right to choose my food. What good is religious freedom if the government has the right to force feed me mind-altering and body-debilitating materials? And I call much of this government-endorsed food material because it's more like material than food. If you want to know what good food is, as a rule of thumb, whatever was available in 1900 is probably okay.

If it has become available since then, it should be suspect. As the local clean food movement moves forward and gets discovered by more and more people who appreciate its quality, integrity, and safety, the industrial food system will push back. It will try to demonize, criminalize, obfuscate, cloud the issues. Let's be happy we're finally attracting attention. I remember when, as a teenager selling chickens, nobody had even heard of the word organic. We've come a long way, baby.

Just to make sure no one can misunderstand my pro-active agenda, let me offer what I think is the moral high ground for policy. I first wrote this for the Virginia Independent Consumers and Farmers Association (VICFA) newsletter in December, 2002.

· Freedom. Period. We believe more freedom is better than less freedom. Those who oppose us should be pushed to defend greater government intrusion and centralized mega-corporate food systems as philosophically superior to personal choice.
· Science. We believe numeric bacterial thresholds, stated as parts per billion or by species, are both detectable and measurable. If these can be met in the kitchen sink or on the backyard clothesline or open air processing facility, who cares? This forces opponents to defend subjective, political, non-scientific regulations, and to agree with the current bureaucracy that empirical thresholds are not desirable.
· Biosecurity. We believe the most accountable food system occurs when neighbors transact with neighbors. Friendship commerce, or what I call relationship marketing, is certainly higher moral ground than bar coded transactions

between distrusting consumers and conglomerates carrying liability insurance war chests.

· Competition. We believe more competition is better than less. Period. Competition pushes all parties to higher levels of achievement and performance. To oppose competition is to deny the foundation of successful athletics and business, not to mention personal achievement and product integrity. Who wants monopolies?

· Bioregional, local, community-based business. We believe that when dollars turn over multiple times within community transactions, it creates healthy economies with stability and opportunity. Opponents must defend larger cities as centers of commerce and capital, with continued economic attrition in 90 percent of the countryside.

· Holism. We believe that a diversified, interconnected infrastructure is more efficient and more nearly mimics nature than one that is fractured, disconnected, and fragmented. The integrated homestead and village concept, where tradesmen like cobblers, harness makers, and butchers lived and worked in close proximity is far superior to mega-processing facilities, single-species CAFOs and every commercial food or fiber item traveling 1,500 miles.

· Biological food systems. We believe food is fundamentally animate, not inanimate. Biological systems carry parameters regarding concentrations, physiology, and habitat beyond which disease, infertility, and pathogens increase to re-create balance. An industrial food system denies these moral and ethical boundaries. Our opponents must applaud the drug-dependent industrial food system as fundamentally better than one with a reverence and awe of living systems.

· Entrepreneurs. We believe that fostering entrepreneurship encourages truly creative ideas and techniques to enter the marketplace. Without market access, the best models in the world will never see the light of day. To be against entrepreneurship is to be against the American dream, and to believe that what we currently have is best, rather than something fundamentally different.

344

- Small business. We believe that small business represents more employment opportunity and is more responsive to cultural desires than large business, which tends to become bureaucratic. Opponents must defend the notion that Tyson is more trustworthy than your neighborhood Mom and Pop, and that Wal-Mart responds to needs rather than creating them.
- Personal trust. We believe that more trust exists between families and neighbors than could ever exist between bureaucrat regulator and common taxpayers. To be sure, plenty of trust seems to exist between regulators and large businesses—collusion, in fact. But, to our opponents, the normal, average person on the street must completely trust bureaucrats and big business with the food supply more than a neighbor. And that is a difficult position to defend.
- Alternatives. We believe that our pluralistic society is stronger for allowing home schooling and private schooling alongside public schooling. Our culture is stronger by letting people buy different kinds of cars or none at all; have health insurance or none at all; build a house or none at all. The freedom to opt out of the mainstream paradigm is the conerstone that preserves the minority view, differentiating between top-down societies and bottom-up societies. Our opponents favor coercing consumers to buy only government-approved food, thereby denying opt-out freedoms. It's Custer vs. Sitting Bull all over again.
- Fair pricing. We believe that requiring $500,000 processing facilities for a farmer to be able to sell one T-bone steak to a neighbor creates price discrimination against small producers and consumers who want to patronize him. With such large overheads, not only do many would-be solutions-based entrepreneurs never start, but those who do must charge exorbitant prices to their customers in order to recoup the inordinately astronomical facility overhead costs. Critics must argue that Farmer Brown processing one T-bone steak is equivalent to Con-Agra processing 5,000 animals per day.

- Modernity. We believe that the information and technology economy have joined rural America, antiquating most of the pathogenic fears and infrastructure problems associated with high-risk food storage and handling from 50 years ago, when many of the food safety regulations were initially written. Rural electrification, stainless steel, front end loaders, polyethylene electric fencing, and infrared scanning technology allow miniaturizing, downsizing, and restructuring of the agricultural economy—in short, modernization. To disagree is to prohibit farmers, consumers, and the food system from joining more modern economic structures.
- Solutions. We believe that a fundamentally restructured food production, processing, and marketing system offers real answers to today's food-borne pathogen epidemic. The problems we have were created by the centralized, dysfunctional industrial system and are inherent within it. The only solutions our critics offer are mandatory irradiation, genetic engineering, and additional government regulation. Thus, our opponents must sell the notion that eating sterilized manure is okay, corporate manipulation of the genetic code is at least as credible as nature's or God's design, and we can indeed have salvation by legislation. Moral high ground, don't you think?

We live in a day of unprecedented solutions because we live in a day of unprecedented problems. Real solutions always threaten the existing problem-creating status quo. As we move forward with an earth-healing, social-healing, farm-healing agenda, we will continue to encounter plenty of naysayers. The system doesn't lack for people who will sell their souls to the entrenched powers. But that's okay, because it just makes our cause greater, and our victory sweeter.

Perhaps it is fitting to end with a couple of quotations.

From Ghandi:

First they ignore you.
Then they laugh at you.
Then they fight you.
Then you win.

From Marquis De Vauvenargues: "The wicked are always surprised to find that the good can be clever."

From Albert Camus: "Integrity has no need of rules."

English Proverb: "The happiness of every country depends upon the character of its people rather than the form of its government."

From Andrew Jackson: "One man with courage makes a majority."

From Sally Ride: "All adventures, especially into new territory, are scary."

From George Bernard Shaw: "I never thought much of the courage of a lion tamer. Inside the cage, he is at least safe from people."

From Arthur C. Custance: "Yet there are some who do 'trouble themselves with such things', and who are still open-minded and who have the possible advantage over others of not knowing enough of what has been said in the past, and are therefore not in a mental straight-jacket as a result. It is possible to know too much of traditional wisdom to be able to learn any more: too much has to be un-learned first. It is amazing to discover what one may NOT see when habit or thought and fear of being counted 'odd' have successfully put blinkers on one's vision."

Psalm 35:19-20: "Let not them that are mine enemies wrongfully rejoice over me: neither let them wink with the eye that hate me without a cause. For they speak not peace: but they devise deceitful matters against them that are quiet in the land."

Amen.

Index

abattoirs....................29, 302

aborted babies.................258

abortions..........................322

adhering dirt.......................81

adrenaline........................149

adulterated........................28

adulteration 29

aerobic compost...............127

agenda...............................76

alcohol.............................155

aluminum.........................309

ambient temperature........147

American Farmland Trust........................227

Amish..............................258

animal control..................325

Appalachian.....................159

artisanal food.............21, 147

atmospheric carbon..........129

auditor.............................289

Australia...........................290

average daily gain............141

avian influenza................264

Avian Influenza Task Force...............................264

B vitamins.......................329

BMPs (Best Management Practices)..........................125

bacon................................66

bacterial protection............82

Bambi..............................321

bandsaws.................173, 176

barbarians........................337

bedding pack....................127

Beers, John A....................74

below-cost-timber sales.....235

Berry, Wendell.................164

biomass............129, 144, 251

biosecurity........................276

bioterrorism.....................231

Black Stallion...................328

blood spots........................82

bone meal........................309

bookshelf.........................115

Boortz, Neal.....................260

bootstrap..........................215

Bovatec...........................243

bovine spongiform
encephalopathy..........16, 308

bribing.............................272

Bromfield, Louis..............137

Brunetti, Jerry..................310

Buchanan, Pat...................24

Buffalo Commons............164

building permit.................218

CAFO's (Concentrated
Animal Feeding
Operations).....................122

CREP (Conservation Reserve
Enhancement
Program..............138

cadavers.............................98

calibrate.............................68

calpain.............................147

camphylobactyer.................16

Canada.....................290, 308

candling.............................83

carbon.....................126, 143

carbon sponge...................127

carbonic acid....................143

carrion.............................308

case...................................59

Chemlawn........................225

Cherokees........................159

Chesapeake Bay
Foundation........................122

chicken manure feeding.....311

child endangerment..........334

chill vats............................44

China................................293

chips.................................236

choice..............................74

chokeholds.......................287

choliforms........................39

circular mills...................176

clear cuts.........................160

cleverness........................312

commercial.......................229

community........................193

competency.......................36

compliance.......................36

composting.................47, 127

composting toilet.............221

comprehensive.................113

conjugated linoleic acid...329

conjunctiva.......................130

connectedness..................182

conspiracy........................101

Consumer Reports...............95

continuous grazing............139

Corporate Welfare..............24

cottage-based food.............22

Courter, J. Carlton.............300

Cousteau, Jacques..............240

cow share..........................333

crab water..........................39

Creekstone Farms..............315

Creutzfeld Jakob disease...102

Curb Market ....…......12

curing................................66

custom...............................53

Dagget, Dan......................164

database............................287

dead zone..........................96

death tax............................259

decentralized....................231

decomposing house..........225

deer....................102

despots....................336

diabetes....................16

dictatorship....................91

disciples....................159

diseases....................276, 296

distiller's grains....................15

dog....................168

Dominion....................322

donations....................342

dreams....................18

drugs....................117

DuPont Circle Farmer's Market....................93

dynasties....................154

eBay....................18

E.coli....................16

easement....................228

eatwild.com....................330

education....................117

egg cases....................87

Eggmobile....................166, 253

electrocuted....................93

embryonic births....................18

eminent domain....................255

Emperor....................242

empirical testing....................100

Environmental Protection Agency....................88

epicormick....................161

equilibrium....................251

equal to....................36

errancy....................290

Esther....................63

evicted....................218

evisceration....................41

excise taxes....................260

exploitation....................200

export....................315

exposure....................235

factory farming...............320

fair tax..........................256

faith.............................241

Fallaice, Linda................268

Fallon, Sally...................312

Farm Animal Certified Humane.......................320

Farm Bills......................283

Farm Bureau Federation....77

farmhands......................205

farmland peservationists...............227

farrowing crates..............327

feather..........................275

Feathernet......................168

fecal cleansing..................85

feedback loop...................19

Fertrell..........................310

feudal serf......................191

fighting cocks..................295

Fishbait Swiss.................342

flat tax..........................117

Food Bank......................210

food choice......................62

Food Security Act.............244

Food System Innovation Zones...........................285

foreign aid......................118

foreigner 280

forward margin................144

fraternity.......................140

fraternity of stupidity.........288

free market.....................154

freedom of choice..............25

freezer beef......................54

frost-free........................138

full-flow valves.................138

gallbladder juice................95

game commission..............165

geese.............................168

genetic diversity.................81

genetically modified organism..........114

Geronimo.........................304

Getty, J. Paul...................159

global warming...............129

Goodlatte, Bob.................296

grass..............................328

grass-based dairying........107

grass-based meat.............329

grass fattening.................314

Green Party......................24

grub tub...........................48

guidance counselor..........198

Guns, Germs, and Steel......16

HACCP (Hazard Analysis and Critical Control Point)..............................101

habitable space.................220

hackers............................303

Hannity, Sean...................209

hatcheries........................326

hawk.................................163

hay shed............................126

herbivores.........................309

high impact........................138

holistic..............................189

Hollandaise 80

homeless............................207

homeschool.......................107

hoof and mouth..................300

Howard, Sir Albert............300

Humane Society of the U.S.....................................321

hydrogen............................143

INS (Immigration and Naturalization Service)......280

illegal aliens.......................258

imbed................................194

immune systems.................16

immunological....................16

imperialist.........................191

impermeable......................34

industrial development bonds...............................254

industrial food...................22

inedible........................45, 82

innovation...........................18

insecticide..............252, 308

integrity.....................73, 152

interest............................216

intra-muscular fat............147

irrigation..........................251

Johanns, Mike.................317

Joseph.............................256

Jungle, The......................102

kelp..................................130

kerf..................................177

Kettle Korn.....................243

knuckleboom....................175

labeling............................46

lagoon.......................122, 125

laissez faire......................154

land grants.......................125

land grant colleges.............283

lard..................................148

leaching............................221

level playing field..............74

liability.............................246

liability insurance.............239

liberating...........................110

libertarian.........................117

license..............................288

local food web...................22

logging.............................119

lysteria..............................16

MSG (monosodium glutamate).........................152

MacAfee, Mark.................268

mad cow...........................308

marten..............................168

mattocks..........................142

Mennonites............258

methane............133

minimalistic............226

minimum wage............201

mob grazing............140

Mollison, Bill............140

mowing............329

muck............137, 251

mucous............272

multi-flora............142

mutated............276

myopic............170

NIMBY (Not In My Back Yard)............182

NIPF (Non-Industrial Private Forests)............235

NRA (National Rifle Association)............20

NRCS (Natural Resources Conservation Service)......122

Nader, Ralph............24

national herd............305

Nations, Allan............86

nitrites............69

non scalable............71

nursing............326

obesity............16

omnivores............309

open pollinated corn..........298

Oprah............286

outsourced............187

overgrazed............140

oxygen............143

PETA............323

packers............303

paradigms............20

Parsons, Stan............134

passion............193

pasteurization............17

pasture pulsing............251

pathogenicity...................231

patriotism................283, 304

peasants..........................154

pen..................................324

Pentagon...........................88

Permaculture...................140

pet..................................170

pet food..........................332

petting zoo......................185

Pharaoh...........................256

pig's feet.........................102

Pilgrim, Bo......................159

pinkeye...........................130

plan...................................59

ponds..............................137

postal................................42

postal worker...................327

press releases..................293

Price, Weston...................312

prime..............................148

prisons............................117

product liability................238

property tax.....................257

prototype...........................18

Pruess, Ellie and Don........108

pubescence......................314

Public Law 90-492.............28

pugging...........................126

Putney, Reid....................108

RFID (Radio Frequency
Identification)...................293

rats.................................323

real estate 257

recovery rate....................177

red button..........................96

regressive income tax........260

religious right...................321

rendering..................267, 309

Republican.......................304

retail establishment............68

retainer............................72

reverse discrimination......201

Right to Farm...................183

riparian zone....139, 140, 222

risk................................239

Rumensin.......................243

ruts................................139

sales tax...........................60

salmonella........................16

Savory, Allan..................164

sawdust..........................217

scalable...........................72

schizophrenia...................40

s c h i z o p h r e n i c
reasoning..................88, 151

science...........................290

scientific method.............131

Scully, Matthew...............322

seasonings......................151

seed money....................137

self-policing......................19

senescence.......................141

set-back..........................255

shackles....................44, 94

silvicultural practice..........128

Sinclair, Upton...........102, 199

Sitting Bull.....................338

slabwood.........................179

slash...............................128

slurry pump.....................129

Social Security.................257

Spanish flu........................16

Specified Risk Materials...313

spinach...........................340

squeegee...........................61

subclinical pneumonia.......272

subcutaneous chips............293

subsidies....................23, 118

suit................................239

sweet potatoes.................211

tallow................................148

tannery.............................186

telecommuting..................193

terrorist............................281

theme parks.....................185

Thomson, A. P. 108

thrift store........................216

Thumper...........................321

top-down............................23

totalitarianism..................288

tours................................184

transparency.....................154

turf war..............................86

Typhoid Mary..................271

tyrant...............................230

Tyson...............................275

under inspection................55

unintended
consequences.........24

VICFA (Virginia Independent
Consumers and Farmers
Association)......................343

vegan...............................321

vegetarians...............321, 324

vehicles............................253

Vermont...........................160

victims...............................75

viewscape.........................257

waiver........................57, 244

walk-in freezer.................282

water-bed pads.................134

Wentling, Roger..............107

western.............................312

white lightning.................334

wineries...........................300

wood chipper...................126

Wythe, George................337

xenophobia......................205

yurt..................................219